Integrating Membrane Treatment in Large Water Utilities

Awwa
Research
Foundation

About the Awwa Research Foundation

The Awwa Research Foundation (AwwaRF) is a member-supported, international, nonprofit organization that sponsors research to enable water utilities, public health agencies, and other professionals to provide safe and affordable drinking water to consumers.

The Foundation's mission is to advance the science of water to improve the quality of life. To achieve this mission, the Foundation sponsors studies on all aspects of drinking water, including supply and resources, treatment, monitoring and analysis, distribution, management, and health effects. Funding for research is provided primarily by subscription payments from approximately 1,000 utilities, consulting firms, and manufacturers in North America and abroad. Additional funding comes from collaborative partnerships with other national and international organizations, allowing for resources to be leveraged, expertise to be shared, and broad-based knowledge to be developed and disseminated. Government funding serves as a third source of research dollars.

From its headquarters in Denver, Colorado, the Foundation's staff directs and supports the efforts of more than 800 volunteers who serve on the board of trustees and various committees. These volunteers represent many facets of the water industry, and contribute their expertise to select and monitor research studies that benefit the entire drinking water community.

The results of research are disseminated through a number of channels, including reports, the Web site, conferences, and periodicals.

For subscribers, the Foundation serves as a cooperative program in which water suppliers unite to pool their resources. By applying Foundation research findings, these water suppliers can save substantial costs and stay on the leading edge of drinking water science and technology. Since its inception, AwwaRF has supplied the water community with more than $300 million in applied research.

More information about the Foundation and how to become a subscriber is available on the Web at **www.awwarf.org**.

Integrating Membrane Treatment in Large Water Utilities

Prepared by:
Jess Brown and **Daniel Hugaboom**
Carollo Engineers, P.C.,
12592 W. Explorer Drive, Suite 200, Boise, Idaho 83713

Sponsored by:
Awwa Research Foundation
6666 West Quincy Avenue, Denver, CO 80235-3098
and
Kansas City Water Services Department
4800 E. 63rd St., Kansas City, MO 64130

Published by:

Awwa Research Foundation
Advancing the Science of Water®

IWA Publishing

DISCLAIMER

This study was funded by the Awwa Research Foundation (AwwaRF). AwwaRF assumes no responsibility for the content of the research study reported in this publication or for the opinions or statements of fact expressed in the report. The mention of trade names for commercial products does not represent or imply the approval or endorsement of AwwaRF. This report is presented solely for informational purposes.

Published by IWA Publishing, Alliance House, 12 Caxton Street, London SW1H 0QS, UK

Telephone: +44 (0) 20 7654 5500; Fax: +44 (0) 20 7654 5555; Email: publications@iwap.co.uk;
Web: **www.iwapublishing.com**

AwwaRF report number 91045F. Originally published by AwwaRF for its subscribers in 2004.
IWA Publishing version published 2005.

ISBN 1 84339 927 X

CONTENTS

TABLES

TABLES (Continued)

FIGURES

FIGURES (Continued)

FOREWORD

The Awwa Research Foundation (AwwaRF) is a nonprofit corporation that is dedicated to the implementation of a research effort to help utilities respond to regulatory requirements and traditional high-priority concerns of the industry. The research agenda is developed through a process of consultation with subscribers and drinking water professionals. Under the umbrella of a Strategic Research Plan, the Research Advisory Council prioritizes the suggested projects based upon current and future needs, applicability, and past work; the recommendations are forwarded to the Board of Trustees for final selection. The foundation also sponsors research projects through the unsolicited proposal process; the Collaborative Research, Research Applications, and Tailored Collaboration programs; and various joint research efforts with organizations such as the U.S. Environmental Protection Agency, the U.S. Bureau of Reclamation, and the Association of California Water Agencies.

This publication is a result of one of these sponsored studies, and it is hoped that its findings will be applied in communities throughout the world. The following report serves not only as a means of communicating the results of the water industry's centralized research program but also as a tool to enlist the further support of the nonmember utilities and individuals.

Projects are managed closely from their inception to the final report by the foundation's staff and large cadre of volunteers who willingly contribute their time and expertise. The foundation serves a planning and management function and awards contracts to other institutions such as water utilities, universities, and engineering firms. The funding for this research effort comes primarily from the Subscription Program, through which water utilities subscribe to the research program and make an annual payment proportionate to the volume of water they deliver and consultants and manufacturers subscribe based on their annual billings. The program offers a cost-effective and fair method for funding research in the public interest.

A broad spectrum of water supply issues is addressed by the foundation's research agenda: resources, treatment and operations, distribution and storage, water quality and analysis, toxicology, economics, and management. The ultimate purpose of the coordinated effort is to assist water suppliers to provide the highest possible quality of water economically and reliably. The true benefits are realized when the results are implemented at the utility level. The Foundation's trustees are pleased to offer this publication as a contribution toward that end.

The development of membrane technologies for potable water industry has led to their widespread regulatory acceptance and use. As the market has matured, a broad range of applications have emerged, including granular media rehabilitation. This project explored the feasibility of integrating membrane technologies into large scale water treatment plants by maximizing reuse of existing filter infrastructure in order to minimize costs.

Walter J. Bishop
Chair, Board of Trustees
Awwa Research Foundation

James F. Manwaring, P.E.
Executive Director
Awwa Research Foundation

ACKNOWLEDGMENTS

The work was made possible through the financial support of AwwaRF and the Kansas City Water Services Department.

There were many individuals and organizations whose cooperation and efforts made it possible to develop the data and analysis presented in this document:

Awwa Research Foundation Project Advisory Committee, Denver, CO., Alice Fulmer, Sun Liang, Scott Freeman, Herve Buisson
Kansas City Water Services Department, Kansas City, MO., John Reddy, Andy Harris, Franklin Pogge
Carollo Engineers, P.C., Boise, ID., Thomas Seacord, Bertrand Alibert, Jason Davis
Carollo Engineers, P.C., Kansas City, MO., Vincent Hart, Thomas Crowley

The authors would also like to recognize the contribution of the following suppliers:
INDECK - Wheeling, IL., Bruce Tait, Brendan Kranzmann, Paul Johnson
Hydranautics - Oceanside, CA., David Faber
Westech/Polymem - Fourquevaux, FR., Jean-Michele Espenan, Olivier Lorain
Westech Engineering - Salt Lake City, UT., Rex Plaizier, Vince Hamilton, Graeme Medworth
Zenon Environmental - Oakville, ON., Ana Calderon, Mark Murphy

EXECUTIVE SUMMARY

Increasing regulatory focus on the microbial quality of drinking water has driven many water utilities to consider advanced treatment technologies such as membranes. Though membrane filtration has been implemented broadly, construction of a new facility is not feasible in some locations due to land availability, desire to reuse existing infrastructure cost implications, or logistical concerns. These constraints are particularly apparent for large water treatment plants (WTPs), where the construction of dedicated membrane filtration facilities and related infrastructure could cost hundreds of millions of dollars and require substantial additional footprint. Previous experimental work conducted at the Kansas City Missouri Water Treatment Plant provided membrane performance and cost data indicating that retrofitting granular media filters with microfiltration (MF) or ultrafiltration (UF) is a cost-competitive alternative for meeting the Long-Term 2 Enhanced Surface Water Treatment Rule (LT2ESWTR) *Cryptosporidium* control regulations. Plant retrofits make use of existing structures, piping, ancillary equipment, and other critical infrastructure. However, retrofitting large WTPs with membranes within the limits of the existing filter gallery is unprecedented and associated implementation issues are undocumented.

RESEARCH OBJECTIVES

The overall objective of this project was to identify and examine the key issues related to retrofitting existing granular media filters with membrane media filters in large water utilities. Specific objectives of the project were to:

- Develop full-scale design parameters for retrofitting the Kansas City Water Services Department (KCWSD) WTP with low-pressure membranes.
- Develop engineering drawings detailing both gravity-fed and pumped membrane retrofit scenarios at KCWSD based on the design parameters developed during pilot testing. This exercise included an analysis of integration issues such as the use of existing infrastructure, ancillary equipment requirements, membrane integrity testing protocols, and operator training requirements.
- Develop cost estimates for retrofitting existing granular media filters with membrane media filters at the KCWSD WTP.

APPROACH

Membrane pilot testing was conducted at KCWSD's 240-million gallon per day (mgd) lime softening plant, which draws water from the Missouri River. The objective of pilot testing was to develop design criteria for retrofitting the granular media filters with membrane media filtration. Three low-pressure membranes were pilot tested at KCWSD: INDECK/Hydranautics UF Hydracap 60, Westech/Polymem UF 120S3, and Zenon ZW-1000. These membranes encompass two system alternatives that may be particularly suitable for retrofit designs: (1) custom designed, pressurized vessel membrane systems, and (2) vacuum-driven immersed membrane systems (Zenon). The pressure driven systems were evaluated under low flux, low feed pressure conditions that would permit their operation within the existing hydraulic grade line at the plant. All three membrane systems were evaluated under high flux, high feed pressure, or permeate vacuum conditions to maximize capacity of the membrane modules and to establish design parameters for a pumped scenario. Three operational criteria were maintained throughout pilot testing: (1) clean-in-place (CIP) interval ≥ 30 days, (2) recovery ≥ 90 percent, and (3) total chlorine concentration added to the backwash ≤ 2 milligrams per liter (mg/L).

The design fluxes and recoveries established during pilot testing were used to develop custom retrofit layouts for the gravity-fed and pumped scenarios. The gravity-fed scenario is a membrane system that works only within the hydraulic grade line of the existing granular media filter galleries, and utilizes the existing infrastructure for backwashing and the existing waste washwater gullet for backwash disposal. The pumped scenario, which is the conventional approach, uses feed water pumps and dedicated membrane backwash stations. Feed pumps permit higher transmembrane pressures (TMPs) and flux rates, which result in fewer required membrane modules, racks and valves, and less interconnecting piping.

CONCLUSIONS

Pilot Testing

- An analysis of the existing hydraulic grade profile at the KCWSD WTP showed that the gravity-fed scenario allowed a maximum available membrane system head loss of 10 feet (4.3 pounds per square inch [psi]), without modifications to downstream hydraulic conveyance systems.
- Gravity-fed flux rates ranged from 30 gallons/ft^2•day (gfd) at 20 degrees Celsius ($^\circ$C) to 41 gfd at 20 $^\circ$C, while minimum operating criteria were met (i.e., CIP interval ≥ 30 days, recovery ≥ 90 percent, and total chlorine concentration added to the backwash ≤ 2 mg/L).
- Pumped scenario flux rates for the pressurized systems ranged from 56 to 87 gfd at 20° C, while minimum operating criteria were met.
- The maximum flux rate for the vacuum-driven immersed system that consistently met the minimum operating criteria was 20 gfd at 20° C. When the median feed water turbidity was greater than 9.9 NTU, the immersed vacuum-driven system did not meet minimum operating criteria at fluxes greater than 20 gfd at 20° C. However, the system operated at a 95 percent recovery, as recommended by the manufacturer, and lowering percent recovery would likely have reduced fouling.
- The addition of 5 mg/L of powdered activated carbon (PAC) to the feed water appeared to lower membrane fouling rates. It is likely that PAC mitigated organic fouling by adsorption and/or that the PAC altered the nature of the membrane surface solids layer, making it more amenable to removal during backwash events.
- Membrane permeability was recovered between runs using hypochlorite and acid cleans, indicating that both organic and inorganic fouling occurred. Caustic cleans decreased membrane permeability.
- A low molecular weight cationic polymer (Polydyne C-308P) was dosed to the common membrane systems feed tank at 1.2 mg/L, (10 times the concentration added to the KCWSD granular media filters during normal operation) over a 3-day period. While fouling resulting from the polymer spiking was readily apparent for the INDECK/Hydranautics membrane, membrane permeability was completely restored

after a hypochlorite and acid cleans were performed respectively. For the Westech/Polymem membrane, no significant increase in TMP was observed during the polymer-spiking test. However, membrane permeability was not fully recovered even after an extensive CIP. The vacuum-driven immersed system was not available for this test due to timing issues.

Retrofit Assessment

- Flux rates and recoveries used to design membrane retrofit layouts for the gravity and pumped scenarios were taken directly from pilot testing results.
- For the gravity-fed approach, membrane racks, modules, and ancillary equipment were required to fit within the existing filter cells. The Westech/Polymem layout met this requirement, while the INDECK/Hydranautics layout required additional footprint for membrane racks and modules. The single feed and permeate ports, a high fiber packing density, and a large active surface area within each module were the differentiating characteristics of the Westech/Polymem module that allowed the system to be retrofitted without requiring additional footprint.
- All three-membrane systems were compact enough to fit within the existing filter gallery under the pumped scenario.
- The number of filter cells required for pumped/feed retrofit layouts varied from 24 to 33 (out of 36 existing filter cells). The most compact designs (highest equivalent hydraulic loading rates) were achieved using modules with the highest fiber packing densities not with those that achieved the highest flux.

Retrofit Costs

- On an average basis, capital costs for the gravity-fed scenario were 9 percent lower than for the pumped scenario, as presented in Figures ES.1 and ES.2, respectively.
- Average membrane equipment costs were estimated to be $0.17/gallons per day (gpd) capacity for both the gravity-fed and pumped scenarios. The increased number of modules and racks required for the gravity-fed scenario was offset by the decreased number of feed and backwash pumps required.
- Due to the costs of additional interconnecting piping and associated valves and fittings to support the larger number of membrane modules, mechanical costs for the gravity-fed scenario were higher than for the pumped scenario.
- Modification of filter cells was minimized in the gravity-fed scenario, which resulted in lower structural costs.
- Electrical costs for the gravity-fed scenario were estimated to be substantially less than for the pumped scenario, as electrical and motor control center upgrades are minimal.
- Membrane replacement intervals drove life cycle costs. Assuming a 5-year membrane replacement interval, membrane replacement costs for the gravity-fed scenario were 64 percent greater than for the pumped scenario. At longer replacement intervals, the energy savings of the gravity-fed scenario offset the increased membrane replacement costs.

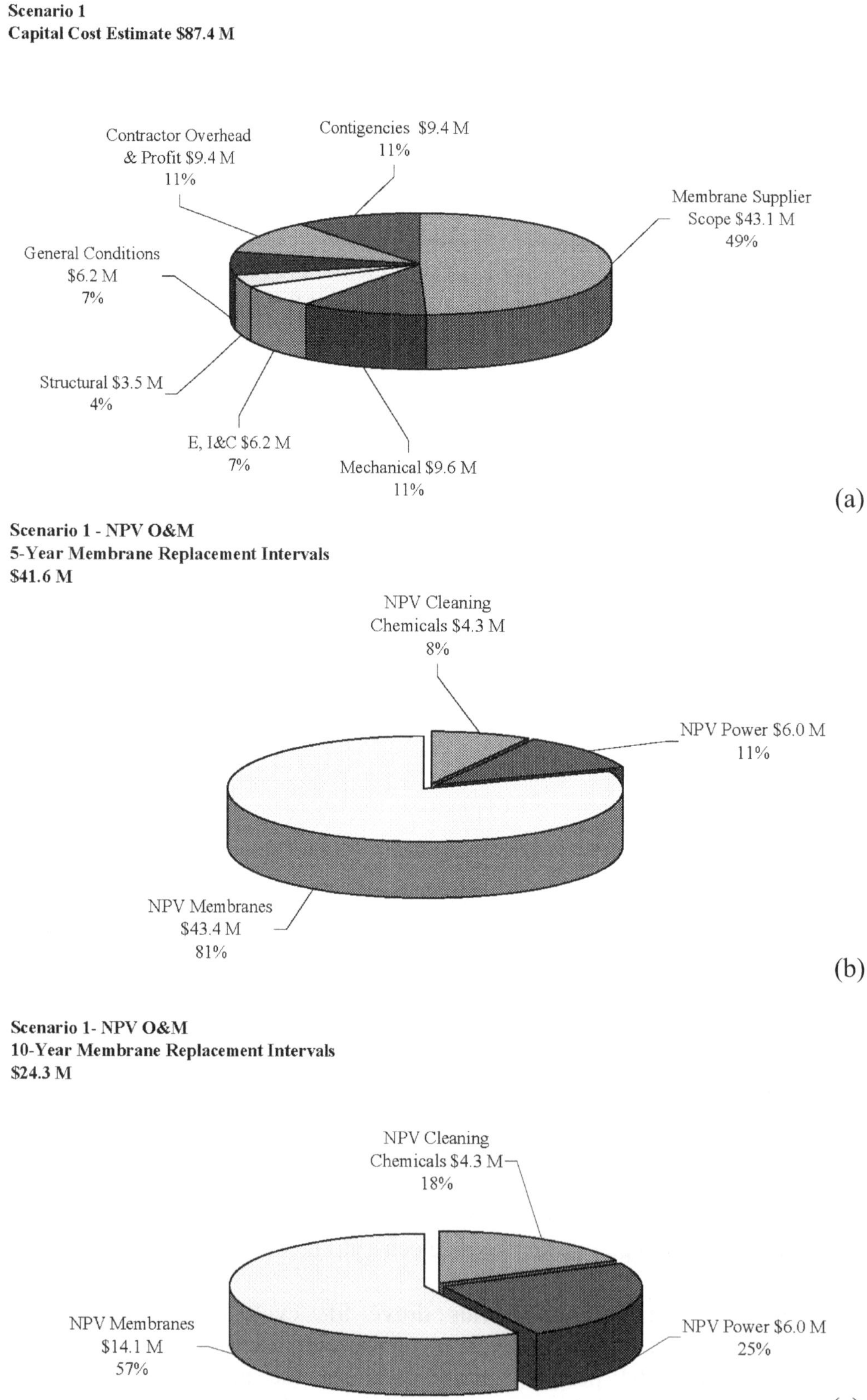

Figure ES.1 Scenario 1 Cost Estimates

Scenario 2
Captial Cost Estimate $94.8 M

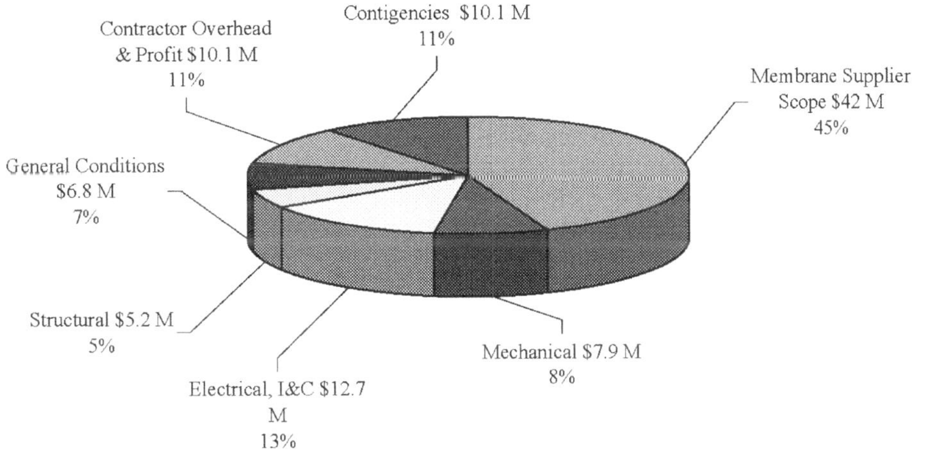

(a)

Scenario 2 - NPV O&M
5-Year Membrane Replacement Intervals
$41.6 M

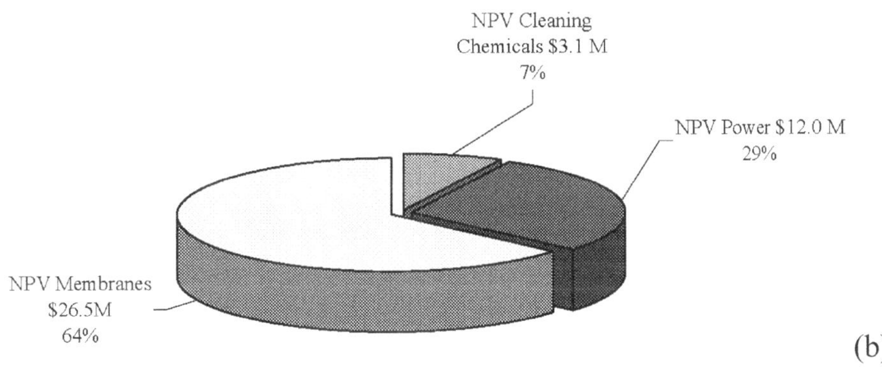

(b)

Senario 2 NPV O&M
10-Year Membrane Replacement Intervals
$22.9 M

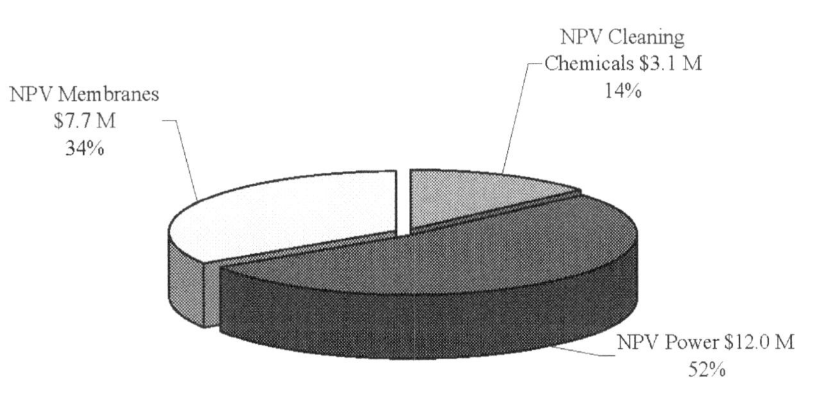

(c)

Figure ES.2 Scenario 2 Cost Estimates

Personnel, Start-Up, and Operations and Maintenance Considerations

- Currently, most states do not require special certification for membrane-based WTP operation.
- Typically, membrane suppliers conduct the majority of operator training sessions on-site once a membrane system has been installed. Training programs specific to the operation of a membrane filtration process may include some or all of the following:
  - Introduction to membrane systems and associated terminology
  - System start-up and shut-down procedures
  - Backwash protocol
  - CIP protocol
  - Integrity testing
  - Fiber repair
  - Module replacement
  - System monitoring and control
  - Trouble shooting
- Special procedures must be implemented to rinse and dispose of the high-biochemical oxygen demand (BOD) glycerin solution in which most membranes are shipped.
- New record keeping protocols may be necessary once a membrane filtration system is in operation. This would include additional water quality parameters as well as integrity test results, fiber repair logs, and cleaning logs.

CHAPTER 1
INTRODUCTION

BACKGROUND

Increasing regulatory focus on the microbial quality of drinking water has driven many water utilities to consider advanced treatment technologies such as low-pressure membranes. As dictated by the Safe Drinking Water Act (SDWA), the Environmental Protection Agency (EPA) has developed a series of rules known as the Microbial and Disinfection By-Product (DBP) Rules. The rules address acute threats to public health from microbial contamination and chronic threats from disinfection residuals and by-products of disinfectants. Currently, large utilities such as KCWSD, must meet filtered water turbidity standards as required by the Interim Enhanced Surface Water Treatment Rule (IESWTR). The rule set initial requirements for removal or chlorine resistant pathogens, such as *Cryptosporidium*. Under this rule, a turbidity limit for combined filter effluent of 0.3 nephelometric turbidity unit (NTU) (95 percent of the time) and maximum limit of 1 NTU were established. In doing so, it is assumed the 2-log *Cryptosporidium* removal is being met. Based on source water *Cryptosporidium* levels, the LT2ESWTR may require higher levels of *Cryptosporidium* removal or inactivation. Among the tools in the "microbial tool box" in the Rules' draft form is membrane filtration.

The Stage 1 DBP Rule set limits for DBPs that have chronic effects on public health. Stage 1 Rules set limits on total trihalomethanes (TTHMs) and haleoacetic acids (HAA_5) of 80 microgram per liter (μg/L) and 60 μg/L, respectively. The Stage 2 Rule will require additional sampling in more points in the distribution system based on an Initial Distribution System Evaluation (IDSE). The rule is expected to be promulgated in July 2005. While membrane filtration does not remove significant levels of DBP precursor compounds, it may reduce the disinfectant dose required to meet disinfection requirements of chlorine susceptible organisms.

MF and UF are now being applied broadly to surface water treatment, combining for more than 1,200-mgd of capacity at over 391 facilities in North America (Chui et al. 2003). In spite of increasing acceptance and application of MF/UF processes, the construction of a new facility is not feasible in some locations due to land availability; desire to reuse existing infrastructure, cost implications, or logistical concerns. These constraints are particularly apparent for large WTPs, where the construction of dedicated membrane filtration facilities and related infrastructure could cost hundreds of millions of dollars and require substantial additional footprint. One alternative to building new membrane facilities is to retrofit WTPs with membrane filtration utilizing existing infrastructure, piping, and ancillary equipment. Additional cost and space savings might be gained if the retrofitted membrane system capitalized on the existing hydraulic grade line of the WTP by using the available gravity head to drive transmembrane flux. However, gravity head across a granular media filtration process may be relatively low, thus necessitating the integration of membranes that can operate at high fluxes under low-head conditions.

Recent work suggests that lime softening pretreatment may be particularly effective at promoting high flux, low head conditions. Lawler and Kweoń (2003) demonstrated that lime softening significantly mitigates UF membrane fouling by removing the hydrophobic and high molecular weight fractions of natural organic matter (NOM). Furthermore, they showed that a settling step following lime softening provides additional reduction in fouling and improvement in transmembrane flux when treating highly turbid water such as the Missouri River. Crozes and Seacord (2001) evaluated the hydraulic performance of MF and UF membranes treating lime softened, settled Missouri River water. Pressure driven polymeric membranes were shown to operate for more than 30 days at recoveries exceeding 90 percent at flux rates of up to 129 gfd at 20°C, with limited fouling. TMP ranged from 14 to 16 psi during the test period, and specific flux loss over the same period was less than 15 percent. A vacuum driven immersed membrane system also was tested at flux rates up to 58.9 gfd at 20°C with 30-day run times. TMPs ranged from 5 to 10 psi.

Results of a worldwide membrane plant survey conducted by Adham et al. (1996) indicate that the economic feasibility of large-capacity MF and UF plants is continually improving through the rapid development of membrane technology. Based on their 1999 pilot work at KCWSD, Crozes and Seacord (2001) developed representative cost curves for new and retrofitted membrane plants. This cost analysis demonstrated that considerable economies of scale could be achieved for large membrane retrofits downstream of lime softening plants. Per gallon capacity costs estimated by Crozes and Seacord (2001), for 10-, 25-, and 100-mgd for completed facilities and installed equipment were $0.47, $0.36, and $0.25, respectively.

Recent developments in membrane technology reflect the growing interest in large-scale membrane filtration facilities. Membranes have been developed that can achieve higher specific fluxes than those tested by Crozes and Seacord (2001). Manufacturers have reported clean water specific fluxes as high as 40 gallons per square foot per day per psi (gfd/psi), compared to 6.5 to 10.1 gfd/psi reported by Crozes and Seacord (2001). Membranes specifically developed for the limited footprints available in retrofits have also been introduced to the market. These membranes have high fiber packing densities, which reduce the number of required membrane modules and do not rely on concentrate recirculation, thereby reducing pumping, piping, footprint, and energy requirements. Despite the flexibility offered by these new technologies and their potential for low footprint and cost requirements, retrofitting filter cells in large WTPs with membranes is unprecedented and associated implementation issues are undocumented.

RESEARCH OBJECTIVES

The overall objective of this project was to identify and examine the key issues related to retrofitting existing granular media filters with membrane media filters in large water utilities. Specific objectives of the project were to:

- Develop full-scale membrane system designs for retrofitting the KCWSD WTP with membranes selected by examining specific characteristics that make them well suited for the application.
- Develop conceptual layouts detailing both gravity-fed and pumped membrane retrofit scenarios at KCWSD based on the design parameters developed during pilot testing. This exercise included an analysis of integration issues such as the use of existing infrastructure, ancillary equipment requirements, membrane integrity testing protocols, and operator training requirements.
- Develop cost estimates for retrofitting existing granular media filters with membrane media filters at the KCWSD WTP.

APPROACH

Membrane pilot testing was conducted at KCWSD's 240-mgd lime softening plant, which draws water from the Missouri River. The objective of pilot testing was to develop design criteria for retrofitting the granular media filters with membrane media filtration. Three low-pressure membranes were pilot-tested at KCWSD: INDECK/Hydranautics UF Hydracap 60, Westech/Polymem UF 120S3, and Zenon ZW-1000. These membranes encompass two system alternatives that may be particularly suitable for retrofit designs: (1) custom designed, pressurized vessel membrane systems and (2) vacuum-driven immersed membrane systems (Zenon). The pressure driven systems were evaluated under low flux, low feed pressure conditions that would permit their operation within the existing hydraulic grade line at the plant. All three-membrane systems were evaluated under high flux; high feed pressure conditions to maximize capacity of the membrane modules and to establish design parameters for a pumped scenario. Three operational criteria were maintained throughout pilot testing: (1) CIP interval ≥ 30 days, (2) recovery ≥ 90 percent, and (3) limits on intermittent chemical cleans to only allow total chlorine concentration added to the backwash ≤ 2 mg/L.

The design fluxes and recoveries established during pilot testing were used to develop custom retrofit layouts for the gravity-fed and pumped scenarios. The gravity-fed scenario is a membrane system that works only within the hydraulic grade line of the existing granular media filter galleries and utilizes the existing infrastructure for backwashing and the existing waste washwater gullet for backwash disposal. The pumped scenario, which is the conventional approach, uses feed water pumps and dedicated membrane backwash stations. Feed pumps permit higher TMPs and flux rates, which result in fewer required membrane modules, racks, and valves and less interconnecting piping.

CHAPTER 2
PILOT TESTING

TESTING SITE

Membrane pilot testing was conducted at KCWSD's 240-mgd lime softening plant, which draws water from the Missouri River. Raw water to the plant is characterized as having high hardness (224/285/170 mg/L as $CaCO_3$)[1], high alkalinity (188/253/130 mg/L as $CaCO_3$)[1] and high turbidity (39/680/10 NTUs)[1]. Typical, yearly-averaged raw water total organic carbon (TOC) concentrations and pH values are near 5 mg/L and 8 mg/L, respectively.

A process schematic illustrating the KCWSD WTP is provided in Figure 2.1. Treatment includes presedimentation, lime softening, stabilization, and sand filtration.

Upstream of the sand filters the following constituents are added:

- Ferric sulfate – coagulant.
- PAC – used for seasonal taste and odor (T&O) episodes and atrazine control.
- Potassium permanganate – oxidant for T&O control and oxidation of manganese.
- Lime – softening agent.
- Carbon Dioxide – pH adjustment.
- Chlorine + ammonia – oxidant for disinfection.
- Sodium hexametaphosphate (SHMP) – used to sequester calcium carbonate through the filters and iron in the distribution system.
- Low molecular weight cationic polymer (Clarifloc C-308P) – filter aid added to the sand filters at 0.12 mg/L. The polymer is also occasionally added to the primary basins to maintain a primary basin effluent turbidity of ≤ 50 NTU.

Stabilized basin effluent (settled water) serves as feed water to the sand filters and was also used to feed the membrane pilot plants. Figures 2.2 and 2.3 illustrate daily settled water turbidity/total chlorine residual and pH/alkalinity values, respectively, during the period over which the membrane pilot testing occurred. A statistical summary of these parameters as measured during the pilots is provided in Table 2.1.

Table 2.1
Statistical summary of settled water pH, alkalinity, turbidity, and chlorine residual (August 1, 2002 to March 31, 2003)

				Percentile		
	Minimum	Average	Maximum	5th	50th	95th
pH	9.0	9.9	10.4	9.4	9.9	10.2
Alkalinity (mg/L)	25	40	63	29	39	53
Turbidity (NTU)	0.2	13	45	1.0	10	38
Total chlorine (mg/L)	1.4	2.3	2.8	2.1	2.4	2.6

[1] (X/Y/Z) = (average/maximum/minimum) values from August 2002 through March 2003.

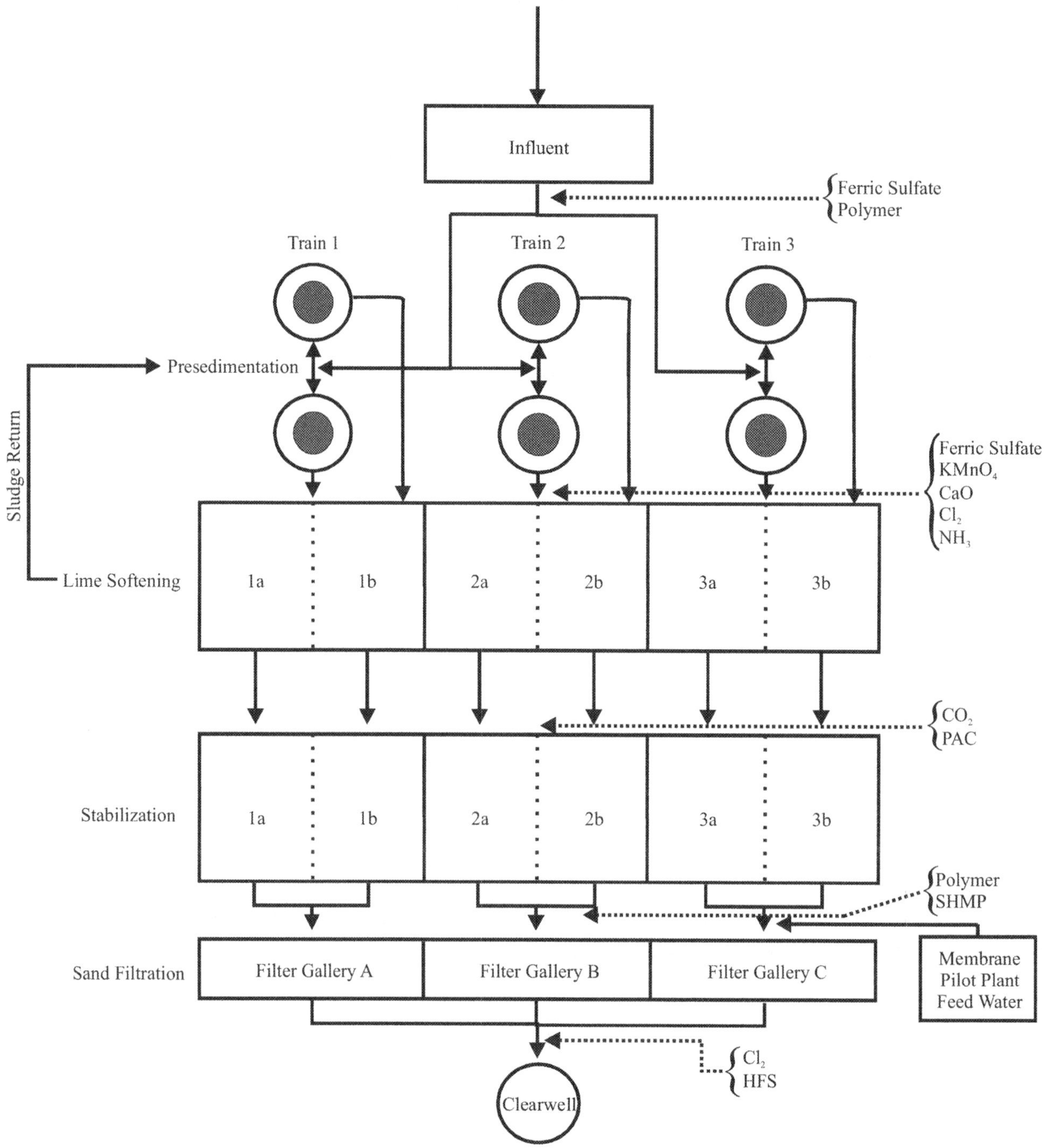

Figure 2.1 KCWSD WTP schematic

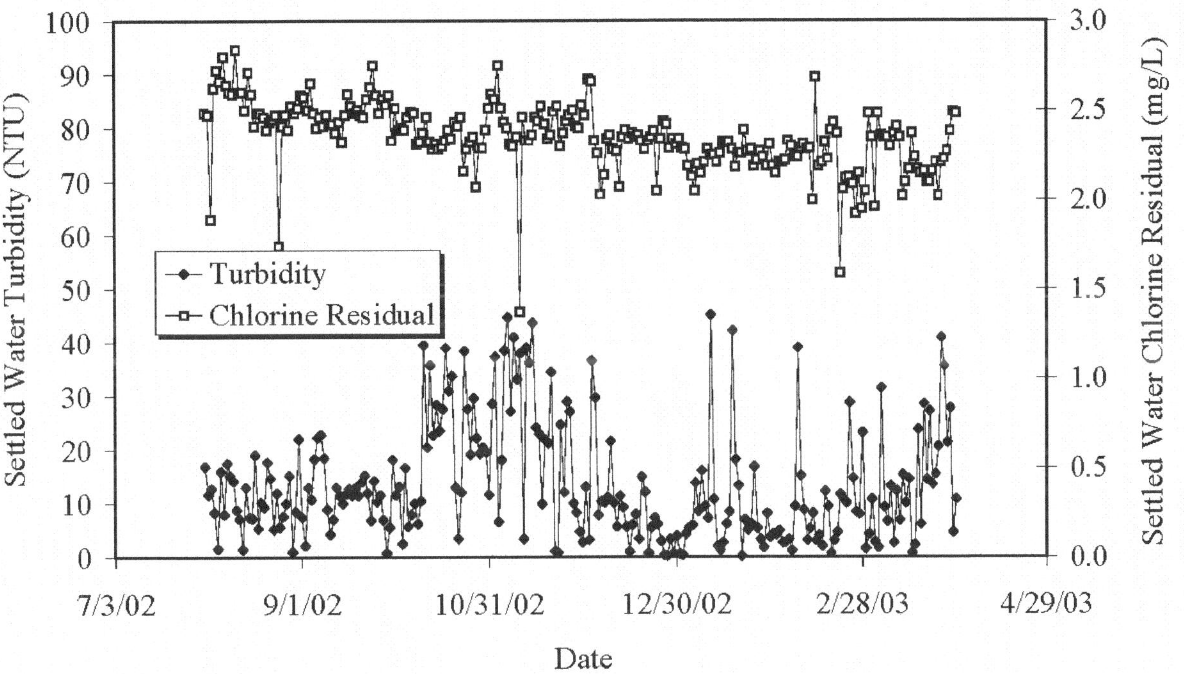

Figure 2.2 Settled water turbidity and chlorine residual

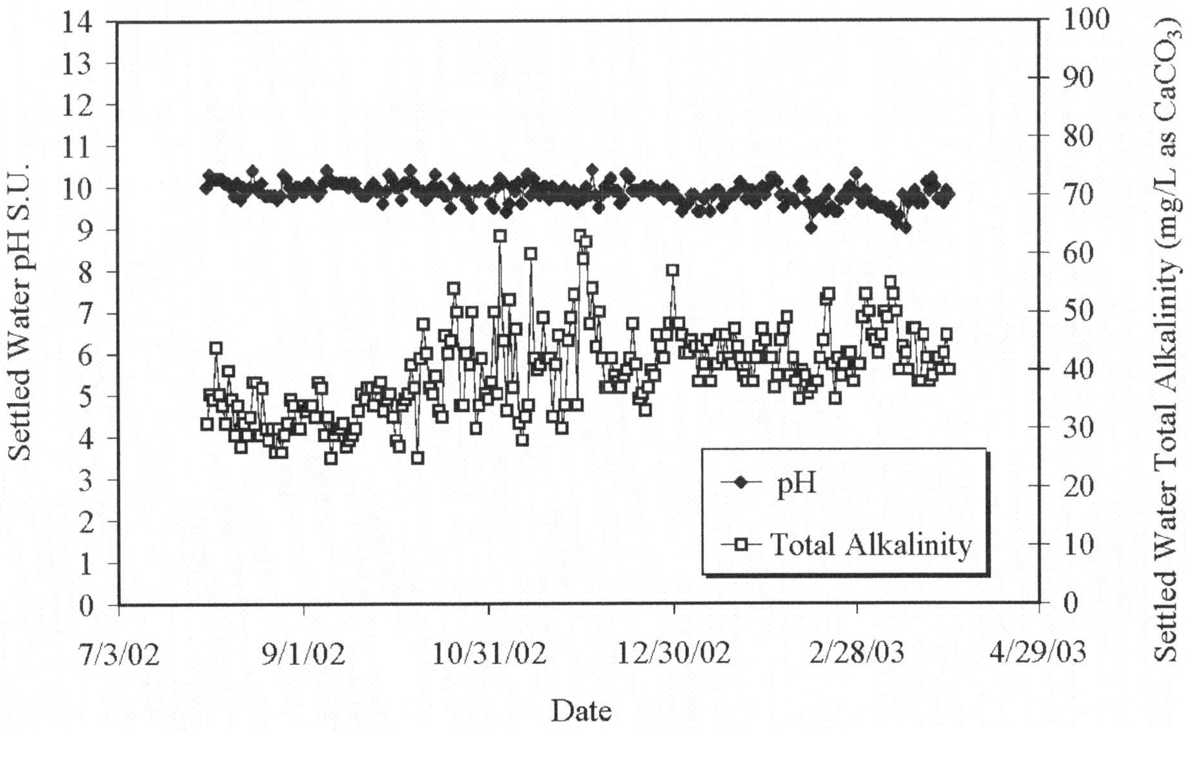

Figure 2.3 Settled water pH and alkalinity

Historically, according to information provided by KCWSD, filtered water turbidity has been dependent on lime dose in the softening process. Membrane filtration, while providing additional disinfection, may also allow KCWSD to decrease lime usage and sludge production.

TESTING PROTOCOL

Objective

The objective of the pilot-testing task was to establish design normalized flux and recovery values, and to characterize and quantify membrane residuals for each membrane system tested.

Operating Criteria

The testing protocol was designed to find an economic optimum for the membrane retrofit and operation by demonstrating high fluxes and recoveries while maintaining lengthy production times and minimized chemical addition. To that end, flux and recovery were optimized for each system, while the following operating criteria were met:

- CIP interval ≥ 30 days.
- Recovery ≥ 90 percent.
- Backwash hypochlorite ≤ 2 mg/L.

Flux and Recovery Optimization

Table 2.2 describes the general approach used to optimize membrane flux and recovery (specific operating conditions for each system are provided in the Results section). All runs were conducted in dead-end mode. Membranes were tested for integrity after initial installation and when filtrate particle counts indicated a possible breach of integrity.

Table 2.2
Flux and recovery optimization

Run designation	Flux @ 20° C	Recovery	Powdered activated carbon
1*	J_1	R_1	N
2*	J_2	R_1	N
3*	J_3	R_1	N
4*	J_4	R_1	N
5†	J_5 (Optimum)	R_1	Y
6‡	J_6 (Optimum)	R_1 or $R_2 > R_1$	N

* Runs 1 through 4 were used to optimize flux, while meeting minimum recovery criteria, under typical feed water conditions.
† Run 5 was used to demonstrate optimized flux and recovery during simulated PAC carry-over conditions.
‡ The purpose of Run 6 was to determine maximum achievable fluxes while operating at recoveries > 90%.

Percent recovery and temperature normalized flux set points for Run 1 were determined during a 1-week preliminary testing period performed by the pilot/membrane suppliers. The goal was to choose operating parameters that would allow for the optimization of flux and recovery over the six-month testing period. Therefore, initial set points were not necessarily identical among membrane systems. For a given run, once normalized flux and recovery were set, each system was operated until one of the following occurred: (1) a target TMP was reached or (2) the run length reached 30 days. Initially, a target TMP of 4.3 psi was set for the pressurized systems so that they could be designed to operate without feed pumps. This TMP represents the existing elevation head at KCWSD. Figure 2.4 illustrates the hydraulic profile for a single filter cell in Gallery C of the KCWSD plant.

Once the 4.3 psi design fluxes and recoveries were established for the pressurized systems, terminal TMPs were then used as targets to find a second set of design fluxes. The purpose of this exercise was to broaden the applicability of pilot testing to large utilities that may have available elevation head greater than 4.3 psi. It also provided design criteria for a retrofitted membrane process that includes feed pumps (or permeate pumps for the vacuum-driven immersed system). If a given run length reached 30 days before the target TMP was reached, the run was stopped and a CIP was performed. Using the membrane performance from the previous run as a guide, the normalized flux was then increased while recovery was held constant. A design flux was established when the actual TMP equaled the target TMP after approximately 30 days of run time. This is illustrated conceptually in Figure 2.5.

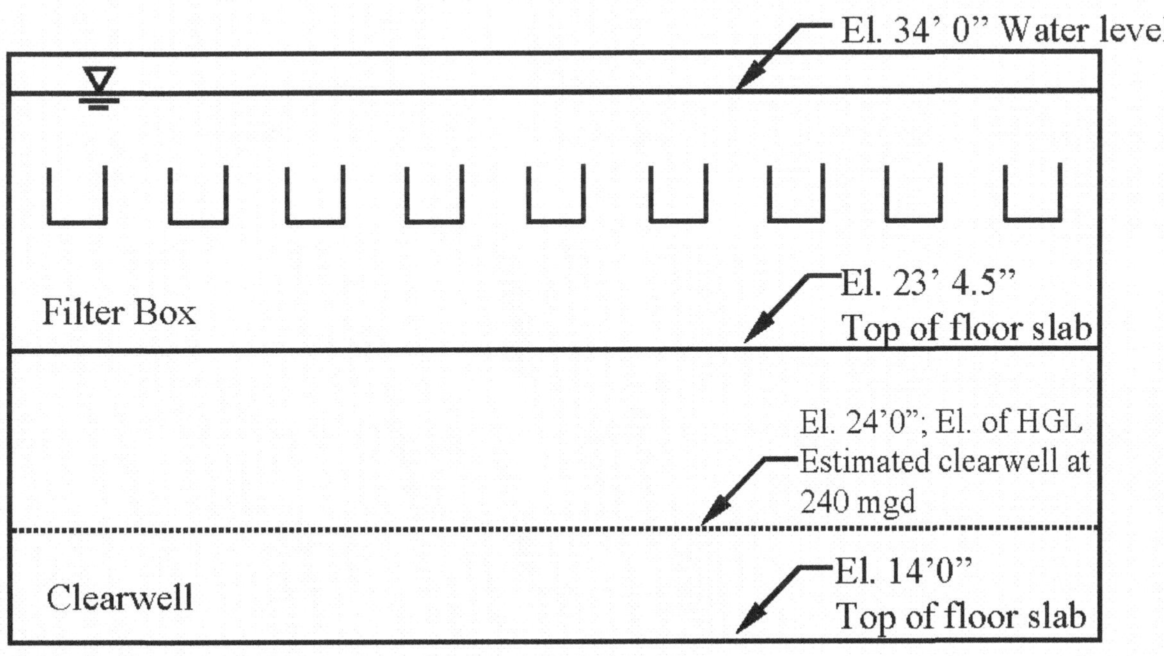

Figure 2.4 Hydraulic grade profile – Filter Gallery C

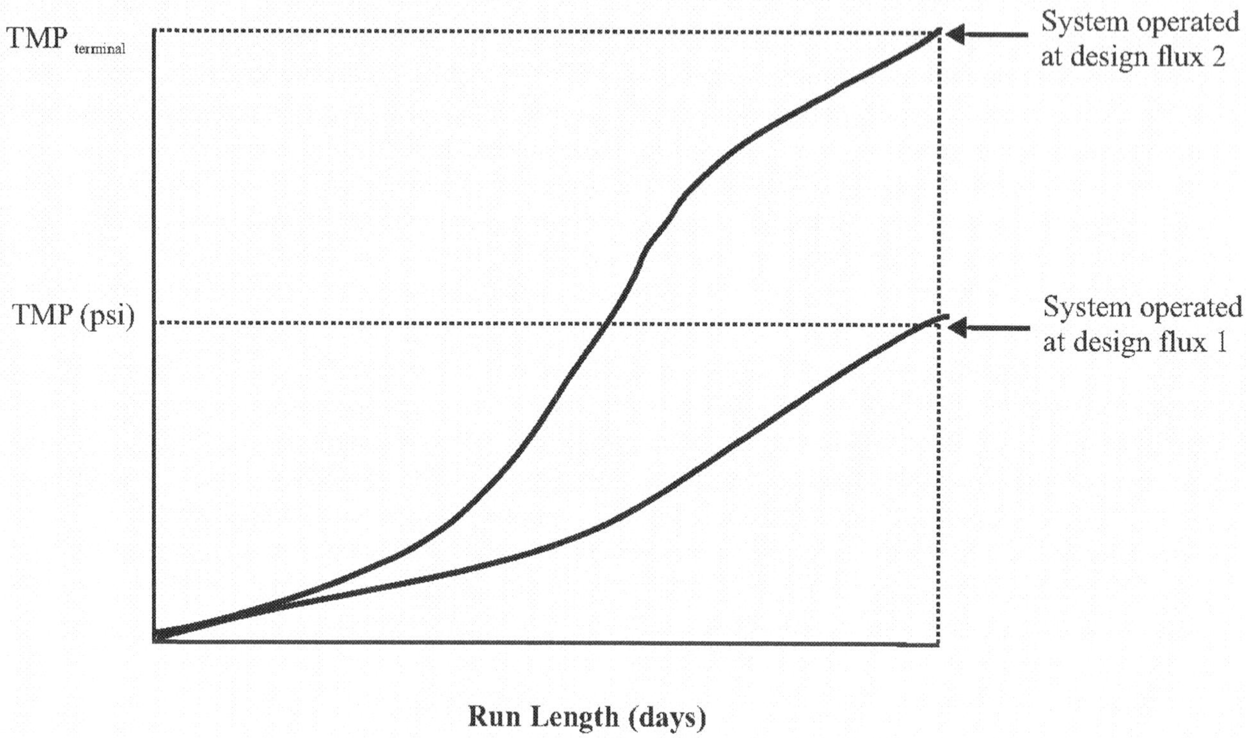

Figure 2.5 Design flux determination

During Run 5, PAC was added to membrane feed water at 5 mg/L to simulate an extreme PAC carry-over event in the filter influent. Optimal fluxes and recoveries, as determined during Runs 1 through 4, were used during Run 5. The purpose of Run 6 was to determine maximum achievable fluxes (using the membrane terminal TMPs) while operating at recoveries $\geq$ 90 percent. At the end of Run 6, a polymer sensitivity test was performed to evaluate the performance of the membranes under a high polymer addition scenario. The feed water to the membrane pilots was spiked with 1.2 mg/L polymer to simulate carry-over from the primary basins. This test was run for 3 days.

IMPACT OF FLUX NORMALIZATION ON DESIGN CRITERIA

Flow rates during the study were calculated based on a target temperature normalize flux and feed water temperature. This was done in order to develop a data set that could be used to calculate system capacity over the range of water temperatures observed at the facility, while pilot testing over a 6-month period.

Much of the pilot testing was conducted at temperatures of less than 20°C hence flux rates at ambient temperature were less than the temperature normalized flux. While normalization accounts for temperature related viscosity effects, it does not account for the impact of mass loading on membrane performance. Flow rates corresponding to higher temperatures could feasibly increase fouling rates during the summer months, due to a higher mass loading rate.

This study covered a 6-month period from September to March with the highest normalized flux occurring in the later months of the study. While it is believed that the data acquired can be used to estimate capacity at different water temperatures, the size of the retrofit (240-mgd) may warrant a longer-term (minimum 1 year) study of hydraulic performance. This period could also be used to verify cleaning procedures over a longer term than covered in this study.

Membrane Integrity Testing

Standard pressure/vacuum decay tests were performed to verify membrane integrity. For each system, the membrane was pressurized according to the manufacturer recommended specification. Once the pressure stabilized, the air was shut-off, and the pressure decay was monitored and checked against the threshold value specified by the manufacturer. Pressure decay tests were performed following initial installation of each membrane module and when particle counters or turbidimeter data suggested degradation of permeate water quality. Permeate particle counts were used as a continuous indirect integrity test throughout pilot testing.

Calculated Process Parameters

Equations 2.1 through 2.10 were used to calculate various process parameters for dead-end UF membrane operation.

Transmembrane Pressure

$$TMP = P_f - P_p \quad \text{(Dead-end)} \tag{2.1}$$

where TMP = Transmembrane pressure, psi
P_f = Feed pressure, psi
P_p = Permeate pressure, psi

Instantaneous Flux

$$J_{T°C} = \frac{Q_p}{A_m} \tag{2.2}$$

where $J_{T°C}$ = Ambient flux of water, gal/ft^2•day
Q_p = Permeate flow, gal/day
A_m = Membrane surface area, ft^2

Specific Flux

$$SF = \frac{J_{T°C}}{TMP} \tag{2.3}$$

where SF = Specific flux, gal/ft^2•day•psi

Temperature Correction Factor

$$TCF = e^{(-0.0239(T_a - 20))} \tag{2.4}$$

where TCF = Temperature correction factor
T_a = Ambient temperature, °C
20 = Normalized temperature, °C

Temperature Normalized Flux

$$J_{20°C} = J_{T°C} * TCF \tag{2.5}$$

where $J_{20°C}$ = Temperature normalized flux of water, gal/ft^2•day

Recovery – Pressurized Systems

$$R = \frac{Q_{pr} * t_{pr} - V_{BW}}{Q_{pr} * t_{pr}} * 100 \tag{2.6}$$

where R = Recovery (%)
Q_{pr} = Production flow, gal/min
t_{pr} = Production time, min
V_{BW} = Volume of permeate water used for backwash

Recovery – Vacuum-Driven Immersed System

$$R_{SS} = \frac{\left[\left(Q_{pr} * t_{pr} - V_{BW} \right) - \left(\frac{V_{PT}}{1440 / t_{pr}} \right) \right]}{Q_{pr} * t_{pr}} * 100 \tag{2.7}$$

where R_{SS} = Recovery (%)
Q_{pr} = Production flow, gal/min
t_{pr} = Production time, min
V_{BW} = Volume of permeate water used for backwash
V_{PT} = Volume of process tank (drained once every 24 hours of production)
1440 = minutes per day

Net Flux

$$J_{NET\ at\ 20C} = J_{Instantaneous} * \frac{J_{NET\ at\ 20C}}{J_{Instantaneous}}$$

$$\frac{J_{NET\ at\ 20C}}{J_{Instantaneous}} = \frac{\left[Q_{PR} * 43200\frac{minutes}{month} * (1 - \%\ Down\ time)\right] - V_{BW} * \frac{no.BW}{month} - V_{CIP} * \frac{no.CIP}{month} - V_{PT} * \frac{no.tank\ drains}{month}}{Q_{PR} * 43200\frac{minutes}{month}}$$

$$\%\ Down\ time = \frac{T_{BW} + T_{IT} + T_{CIP} + T_{TD}}{43200\frac{minutes}{month}}$$

where Q_{PR} = Production flow, gal/min (2.8-2.10)

V_{BW} = Volume of permeate water used for backwash
V_{PT} = Volume of process tank (drained once daily)
V_{CIP} = Volume of CIP solution for 1 CIP
T_{BW} = Total monthly backwash duration
T_{IT} = Total monthly integrity test duration
T_{CIP} = Total monthly CIP duration
T_{TD} = Total monthly tank drain duration
J_{NET} = Net flux
$J_{Instantaneous}$ = Instantaneous (production) flux

Analytical Methods

Water quality data were gathered using on-site bench-top equipment, on-line instruments, and laboratory analyses. KCWSD performed all laboratory analyses except for algae enumeration, which was performed by Montgomery-Watson Harza Laboratories in Pasadena, California. For laboratory analyses, samples were taken in containers supplied by the laboratory and were stored at 4°C until analyzed. Monthly duplicate samples and 5 percent blind samples were taken to measure data reproducibility.

pH

pH was measured using Standard Method 4500-H+ B. A 3-point calibration of the pH meter was performed daily.

Temperature

Temperature was measured daily in accordance with Standard Method 2550 B to check in-line temperature transmitters on the pilot plants.

Turbidity

Feed and permeate water in-line turbidity measurements were logged continuously on the pilot plant using Hach 1720D low-range process turbidimeters. The feed and permeate turbidimeters were calibrated after each run using primary calibration standards purchased from the manufacturer. Lenses were cleaned every two weeks according to the manufacturer's instructions. Turbidimeter flow rates were checked daily to ensure that they are within the range required by the manufacturer. Daily bench-top turbidity analyses were performed using a Hach 2100 AN turbidimeter to verify in-line turbidity measurements. The turbidimeter was checked daily against primary turbidity standards (0, 20, 100, and 800 NTU) and calibration was performed if any measurement deviated from a standard by more than 5 percent. Turbidity was also measured daily in the KCWSD laboratory using Standard Method 2130 B.

Total Suspended Solids

KCWSD laboratory measured total suspended solids according to Standard Method 2540 D.

Total Alkalinity

KCWSD laboratory measured total alkalinity according to Standard Method 2320 B.

Total Hardness

KCWSD laboratory measured total hardness according to EPA Method 200.7 and Standard Method 2340 B.

Organic Carbon

KCWSD laboratory measured TOC and dissolved organic carbon (DOC) according to EPA Method 415.1.

UV_{254} Absorbance

KCWSD laboratory measured UV_{254} absorbance according to Standard Method 5910 B.

Total Dissolved Solids

KCWSD laboratory measured total dissolved solids according to Standard Method 2540 C.

Total Chlorine

Total chlorine concentration in membrane backwash water was measured daily using an on-site, bench HACH DR700 spectrophotometer according to HACH Method 8167 (equivalent to Standard Method 4500-Cl G). KCWSD laboratory total chlorine analyses were performed according to Standard Method 4500-Cl D.

Microbial Parameters

KCWSD laboratory measured heterotrophic plates counts (HPCs) and total coliform counts according to Standard Methods 9215 B and 9221 D, respectively. Montgomery Watson Harza Laboratories (Pasadena, CA) enumerated algae according to Standard Method 10200 F.

Particle Counts

Feed and permeate water in-line particle count measurements were logged continuously on each pilot plant using Met One PCX (Westech/Polymem), Hach 2200 PCX (INDECK/Hydranautics), and Chemtrac Lasertrac PC 2440D (Zenon) particle counters. All particle counters underwent regular maintenance (cleaning and flow adjustment) and were operated under current manufacturer calibration certificates.

Disinfection By-Products

KCWSD laboratory measured TTHM and HAA_5 concentrations according to EPA Methods 502.2 and 552.2, respectively.

SYSTEM DESCRIPTIONS AND PILOT TESTING RESULTS

INDECK/Hydranautics

System Description

The treatment steps in the INDECK/Hydranautics system process are shown on the Process Flow Diagrams (PFD) in Figures 2.6 and 2.7 for dead-end filtration mode and backwash mode, respectively. (The system can also operate in cross-flow filtration mode; however, performance in this mode was not evaluated due to footprint constraints and incompatibility with the available driving head.) A brief description of operating modes and other critical functions is presented below.

Membrane module. The characteristics of the HydraCAP 60 membrane module used in this study are summarized in Table 2.3. The module is 60 inches long by 9 inches in diameter with 500 ft^2 of active membrane surface area (AMSA). Clean membrane permeability observed at the site was approximately 15 gfd/psi at 20°C. The membrane fiber and module construction is compatible with the elevated feed water pH (> 10) and oxidant concentrations at the plant (2 to 3 mg/L total chlorine). Module materials also allow the membrane to be cleaned at low pHs (< 3) required to remove scale and with elevated hypochlorite concentrations during CIP events. The membrane operates in an inside to outside (I/O) filtration mode, and is backwashed from outside to in (O/I).

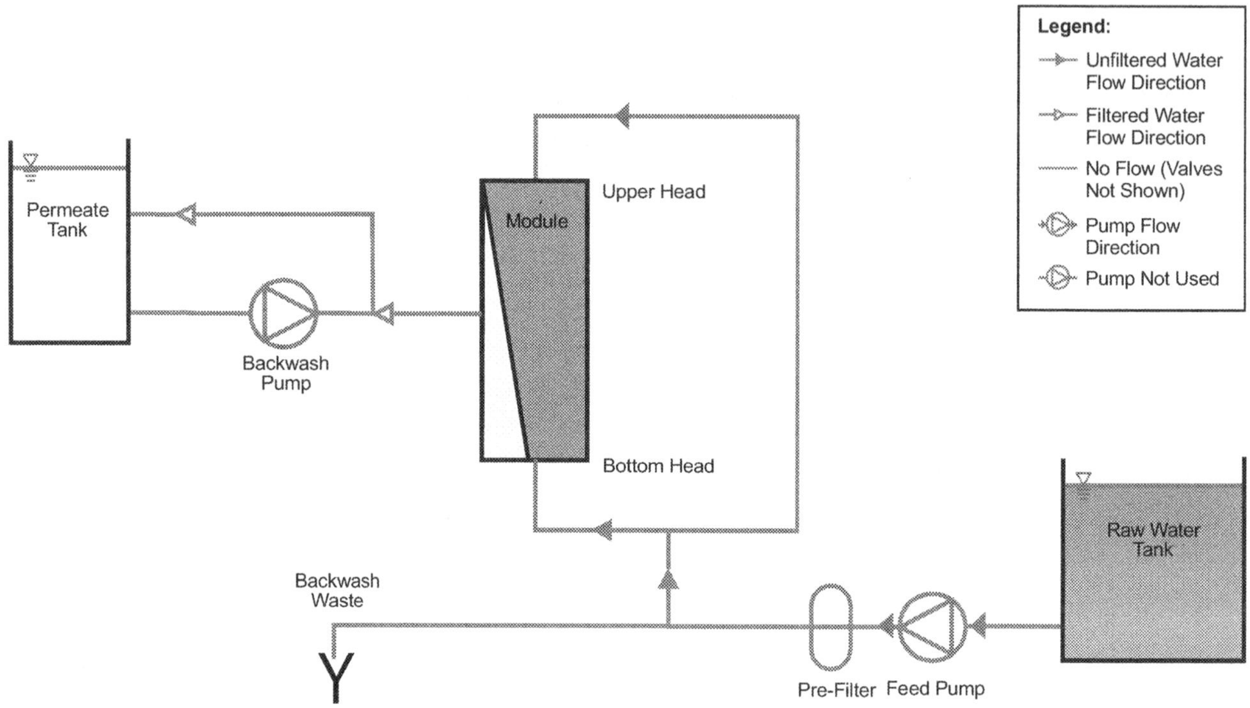

Figure 2.6 INDECK/Hydranautics system PFD (filtration)

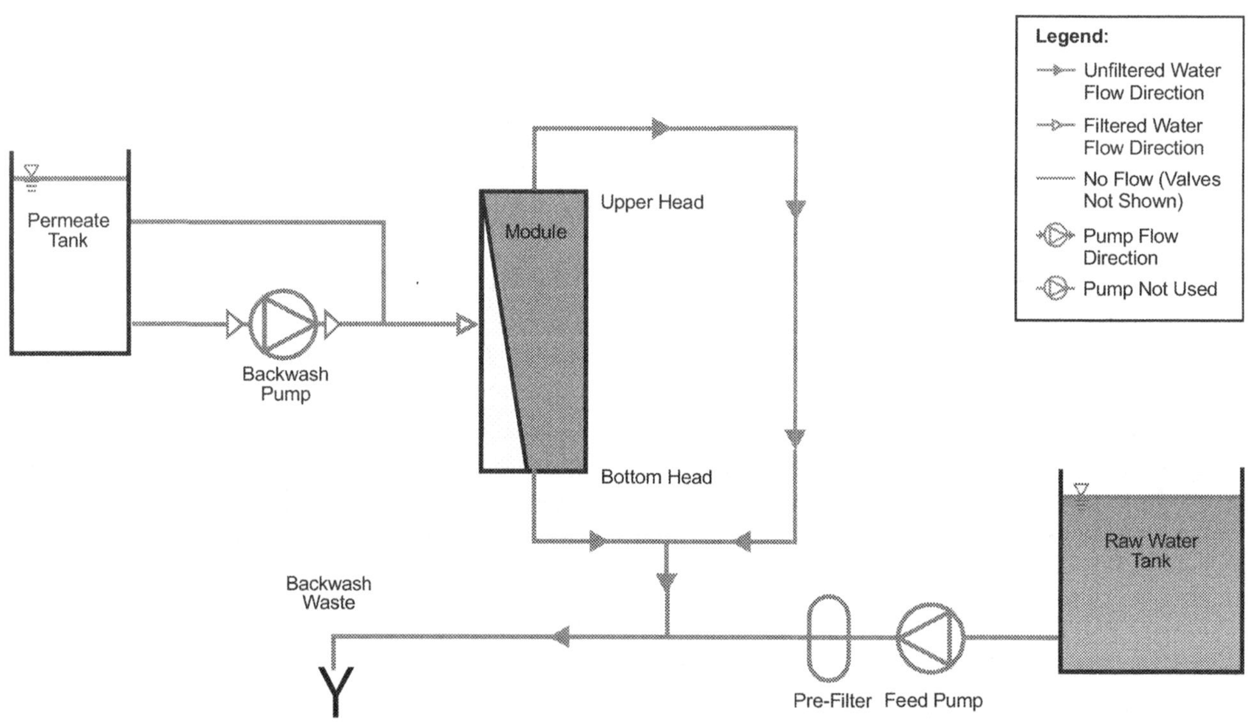

Figure 2.7 INDECK/Hydranautics system PFD (backwash)

Table 2.3
INDECK/Hydranautics membrane module characteristics

	INDECK/Hydranautics
Membrane type	UF
Model	HydraCAP 60
Module dimensions	H=60"; Diameter=9"
Maximum flux	85 gfd
Maximum TMP	22 psi
Area per module (ft^2)	500
Flow direction	Inside-Out
Inside diameter (mm)	0.8
Nominal pore size (Micron)	0.01
Molecular weight cutoff (Daltons)	150,000
Membrane material	Hydrophilic Polyether Sulfone
pH range	2 – 13
Hypochlorite tolerance	5 mg/L constant 200 mg/L instantaneous 200,000 mg/L Hrs (Lifetime)

Source: Adapted from INDECK/Hydranautics, 2002.

Full-scale membrane modules are mounted vertically on racks and hydraulically connected to header pipes. The number of modules per rack is based on site-specific factors including required capacity and flux. In most applications, feed, backwash, and recirculation pumps provide the energy required to perform filtration, backwash, and cleaning operations.

Filtration. Prior to reaching the membrane, the raw water is pre-filtered (150 micrometer [µm]) to eliminate particles that could block the entrance of the hollow fibers. The feed water enters the membrane module through one of two feed water connections located at the ends of the module. Feed water direction alternates between top to bottom and bottom to top in order to uniformly foul the membrane. Feed water enters the membrane fiber from the inside of the lumen, passes across the membrane surface, and is collected in a central module core tube. Core tubes are tied into the permeate header via a single external connection.

Backwash. During filtration, solids accumulate on the membrane surface. A short cleaning, called the backwash, is periodically performed with filtered water to restore the membrane performance (Figure 2.7).

Typically, backwash occurs every 20 to 60 minutes of membrane operation for a duration of approximately 60 seconds. Water is delivered to the outside of the membrane fiber during a backwash, which forces accumulated solids out through the feed side (inside) of the membrane fiber. The backwash flow rate is generally 2 to 4 times the filtration flow rate. Backwash water is discharged from the system through the feed water piping. In some applications, backwash water is amended with hypochlorite (up to 50 mg/L) or with acidic or caustic solutions to mitigate membrane fouling.

Backwash intervals and chemical dosing strategies are set on an application-specific basis to meet cleaning and recovery goals. The backwash procedure used for this study is described herein. Raw water entered one feed port and exited the feed port at the opposite side of the module with the permeate valve closed at 45 gpm (130 gfd) for 10 seconds. Permeate water was then backwashed through the membrane in a counter-current (opposite to the filtration direction) mode at 65 gpm (187 gfd) for 20 seconds, which was followed by a co-current backwash (backwash waste flowing the same direction as the feed water during filtration) at 65 gpm (187 gfd) for 20 seconds. Hypochlorite was added at 2 to 3 mg/L during each 20-second backwash period. The backwash sequence was followed by a 60-second soak period, which was followed by a 10-second rinse cycle. No air scour was used.

Clean-in-place. Backwashes help to reduce the TMP build-up but do not restore the full capacity of the membranes. The membrane must receive a chemical cleaning when the terminal TMP for the run is met. The chemical cleaning removes foulants from the membrane fibers not removed by backwashing alone. Membrane cleaning is typically performed by taking the system out of service and recirculating (in the direction of filtration –I/O) a low pH (pH = 2) solution, high pH (pH = 12) solution, and/or sodium hypochlorite (200 mg/L) as required.

Integrity testing. Membrane integrity is monitored with direct and indirect methods. Indirect methods included turbidity monitoring and particle counting. These parameters are measured with continuous monitoring devices, without the system being taken out of service. The system uses a pressure decay test (PDT) as the direct integrity test (IT), which requires the membrane system to be taken out of service. During the test, the feed side of the membrane fibers are pressurized, and the permeate side is left open to the atmosphere. Pressure in the feed side is monitored, and a decay rate is calculated. Pressure decay in excess of that associated with transmembrane air diffusion indicates a breach of membrane integrity. The IT requires approximately 10 minutes, and results in a total system downtime of 20 to 30 minutes. IT results are listed in Table 2.4.

Pilot Testing Results

Hydraulic performance. Table 2.5 lists the production set points for the INDECK/Hydranatics pilot plant (INDECK/Hydranautics membrane) during Runs 1 though 6. Recovery was set at 90 percent during PAC testing (Run 5), but was maintained at 93 percent for all other runs. During Runs 1 and 2, the INDECK/Hydranautics pilot plant utilized an ambient flux set point instead of a normalized flux set point. This was due to the manner in which the pilot plant was designed. In order to maintain a constant normalized flux, the operator must vary the flow rate daily based on changing temperatures. However, it was decided to convert to a constant normalized flux set point for Runs 3 through 6.

Table 2.4
INDECK/Hydranautics integrity test results

Date	Criteria	Result
09/06/2002	0.18 psi/min	Pass
10/08/2002	0.18 psi/min	Pass
10/08/2002	0.18 psi/min	Pass

Table 2.5
Production set points – INDECK/Hydranautics

Run	Flux @ 20° C (gfd)	Recovery (%)	PAC	Polymer
1	41.0 (ave.)	93	N	N
2	58.0 (ave.)	93	N	N
3	65.0	93	N	N
4	78.0	93	N	N
5	58.0	90	Y*	N
6	87.0	93	N	Y†

*Dosed at 5 mg/L
† Dosed at 1.2 mg/L during final 1.5 days of Run 6

Figure 2.8 (Run 1, INDECK/Hydranautics) illustrates the standard method in which feed water temperature and hydraulic performance data were recorded and plotted during pilot testing. Similar graphs for Runs 2 through 5 are provided in Appendix A (Figures A.1 through A.4). The gap in data between September 10 and September 19 and the smaller gap between November 22 and 23 were due to a loss of communication between the PLC and the pilot plant laptop. This communication problem was fixed for subsequent runs.

A summary of run lengths, temperature trends, and hydraulic performance data for all six INDECK/Hydranautics runs is provided in Table 2.6. Average feed water temperatures were warm during the first run (23°C) but decreased substantially during Run 2 (13°C). Average temperatures dropped to single digits during Runs 3 through 6. A summary of the INDECK/Hydranautics hydraulic performance (non-PAC runs) is presented graphically in Figure 2.9.

Table 2.6
Summary of INDECK/Hydranautics hydraulic performance

Run #	Dates (2002 - 2003)	Length of run (days)	Temperature (range, average; °C)	Recovery (%)	Flux at 20°C (gfd)	Net Flux @ 20°C (gfd)	TMP (@ start of run, @ end of run; psi)	Specific Flux @ 20°C (@ start of run, @ end of run; gfd/psi)
1	9/5-10/5	31	17-28, 23	93	41.0	39.4	3.1, 3.9	13.3, 11.3
2	10/10-11/7	27	8-19, 13	93	58.0	55.2	4.7, 8.8	10.7, 8.6
3	11/8-12/7	30	5-10, 7	93	65.0	62.0	5.6, 10.1	11.6, 6.4
4	12/7-1/6	31	3-8, 4	93	78.0	73.7	9.0, 12.9	8.6, 5.3
5	1/23-2/21	30	3-7, 6	90	58.0	54.8	4.1, 4.2	13.7, 13.4
6	2/25-3/30	34	4-12, 8	93	87.0	84.3	5.7, 12.0	15.7, 6.3

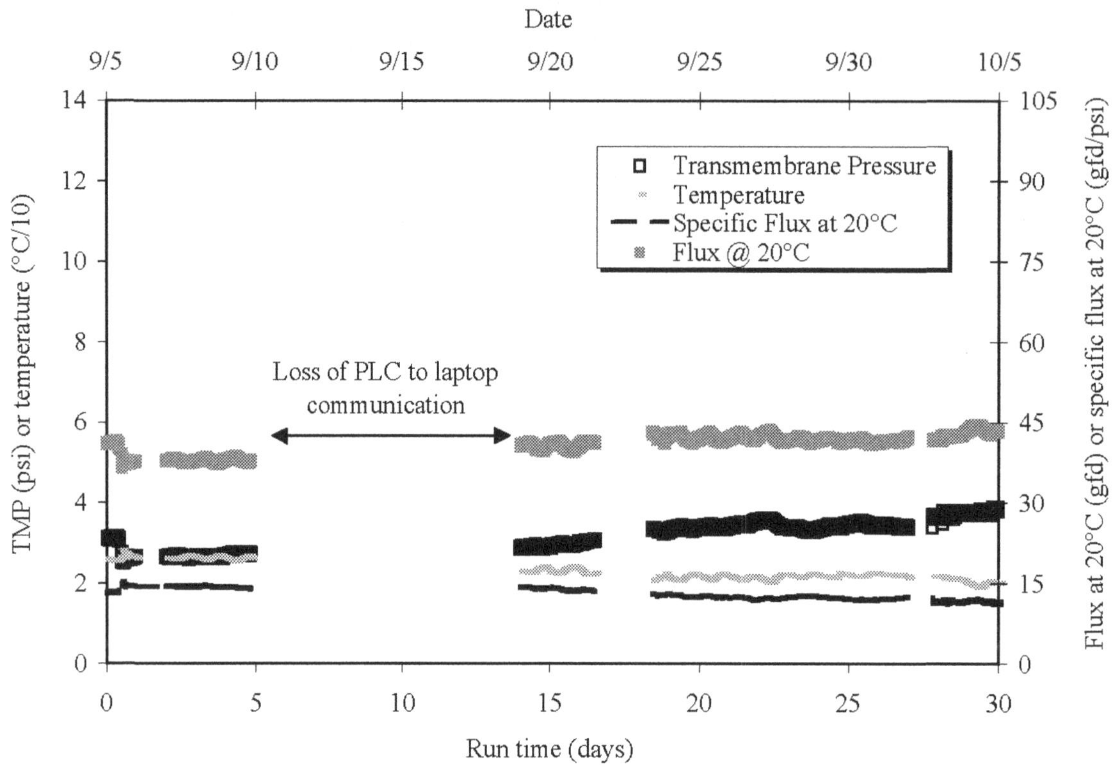

Figure 2.8 Run 1 INDECK/Hydranautics hydraulic performance (operating conditions: average flux @ 20° C = 41.0 gfd, backwash interval = 50 minutes, recovery = 93 %)

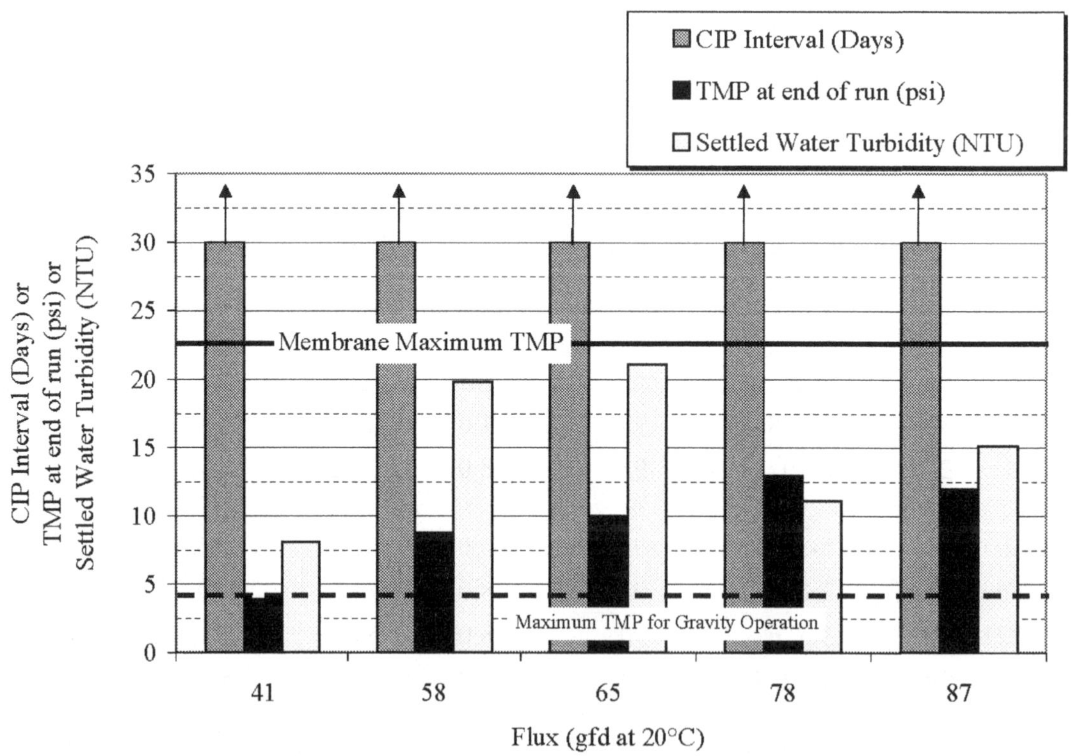

Figure 2.9 Hydraulic performance summary- INDECK/Hydranautics

With an average normalized flux of 41.0 gfd, Run 1 resulted in a final TMP of only 3.9 psi, less than the target 4.3 psi TMP for the gravity-fed scenario. However, increased normalized flux during Run 2 (58.0 gfd, on average) resulted in a maximum TMP of 8.8 psi. Therefore, a recovery of 93 percent and a normalized flux of 41 gfd were set as design parameters for the INDECK/Hydranautics gravity-fed scenario.

Runs 3, 4, and 6 were used to develop design parameters for the pumped feed scenario. A 65.0 gfd normalized flux set point for Run 3 resulted in a maximum TMP of 10.1 psi, substantially lower than the membrane's 30 psi terminal TMP. Therefore, the normalized flux set point was increased to 78.0 gfd for Run 4. Since the TMP had reached only 12.9 psi by the end of the 31-day Run 4, the normalized flux for Run 6 was set at 87.0 gfd, just above the maximum manufacturer-recommended operating flux of 85.0 gfd. Since TMP increased from 5.7 to 12.0 psi, a recovery of 93 percent and a normalized flux of 85 gfd were chosen as design parameters for the INDECK/Hydranautics' pumped-fed scenario.

With the exception of turbidity, feed water quality varied over a narrow range. Figure 2.9 shows the relationship between flux, turbidity, cleaning interval, and final TMP for non-PAC runs. The last 3 days of operation for Run 6 were omitted to remove the impact of excess polymer dosing. While final TMP increased with increasing flux, all runs resulted in terminal TMP less than the maximum as specified by the manufacturer. Though median turbidity for individual runs ranged from 8 to 21 NTU, there is no correlation evident between median turbidity and final TMP for this membrane.

The 5 mg/L of PAC added to the feed water during Run 5 did not hinder membrane performance, as TMP increased by only 0.1 psi and specific normalized flux values (i.e., permeability) decreased by a total of 0.3 gfd/psi over the course of the run. This performance was actually better than that observed during Run 2, when identical production set points were used. These results suggest that the PAC may have been removing organic matter that was fouling the membrane during non-PAC runs. This phenomena has been observed by other researchers (Lâiné et al. 1990, Adham et al. 1991) who have hypothesized that the larger particles present in PAC decrease the hydraulic resistance of the fouling layer. Other researchers have hypothesized that these same particles may enhance particle back transport and provide a surface scouring function (Wiesner and Lâiné 1996). Crozes and Seacord (2001) observed that adding PAC at high concentrations increased specific flux with an I/O polysulfone membrane with the same feed water tested in this study.

At the end of Run 6, 1.2 mg/L of low molecular weight, cationic polymer (Clarifloc C308P) was dosed to the feed water continuously for 1.5 days. During this period, TMP increased from 12.0 to 14.0 psi, causing membrane permeability to decrease from 7.2 to 6.3 gfd/psi (Figure 2.10). Deterioration of permeability was likely the direct result of organic fouling caused by the high polymer concentration dosed to the feed water. Details related to the recovery of permeability after each run are provided in the *Membrane Cleaning Results* section.

During a 2-day period within the third INDECK/Hydranautics run, data logging frequencies were increased five-fold to once per minute. This approach essentially allowed the data to serve as a "zoom" for hydraulic performance, which revealed detailed water quality and hydraulic phenomena associated with backwashing the membrane. While permeate turbidity was unaffected by backwashing, permeate particle counts showed a definite trend related to each backwash interval (Figure 2.11). Immediately following a backwash, permeate particle counts

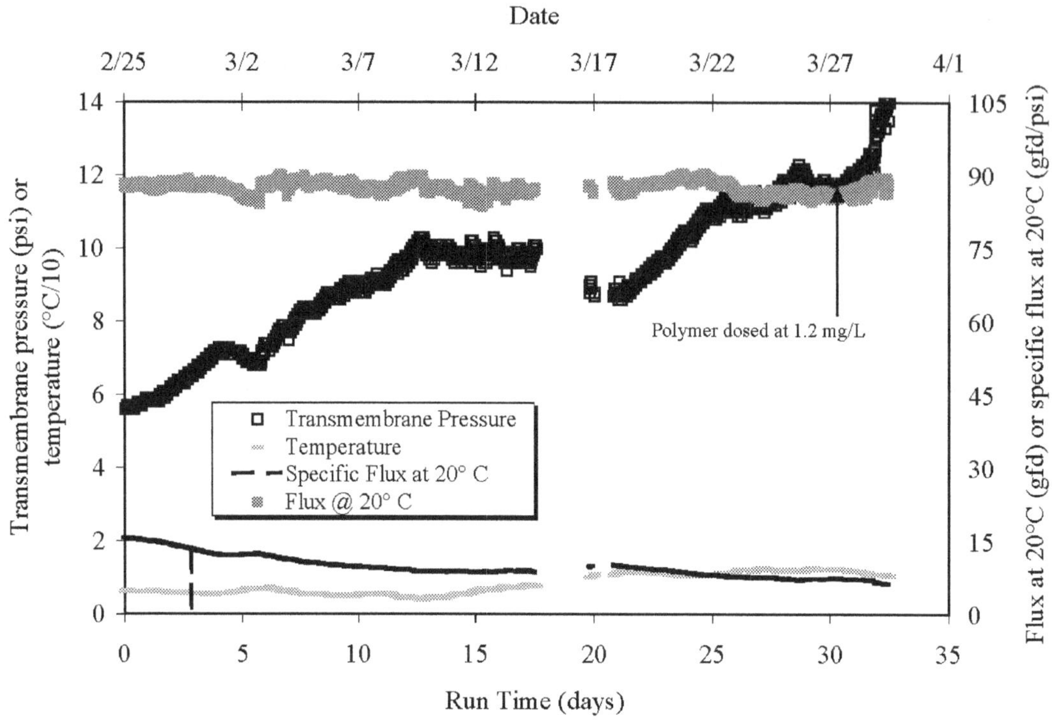

Figure 2.10 Run 6 INDECK/Hydranautics hydraulic performance (operating conditions: flux @ 20° C = 87.0 gfd, backwash interval = 31 minutes, recovery = 93 %)

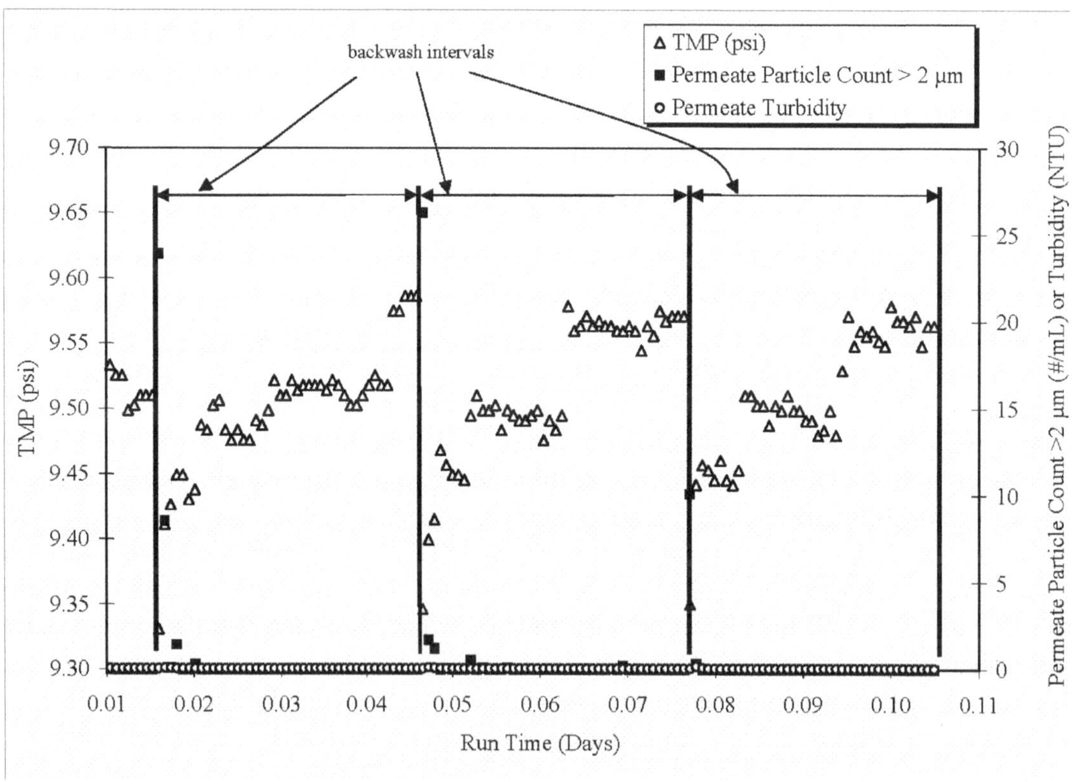

Figure 2.11 Run 3 INDECK/Hydranautics hydraulic performance with zoom (operating conditions: flux @ 20° C = 65.0 gfd, backwash interval = 42 minutes, recovery = 93%)

increased from approximately zero to over 20 counts/mL. Within a few minutes following a backwash, permeate particle counts dropped back to near zero for the remainder of the backwash interval. A trend in TMP was also observed. TMPs were lowest immediately following a backwash and increased by approximately 0.2 psi over the course of a backwash interval. This trend reflects the reversible fouling of the membrane during production, and subsequent removal of the fouling during a backwash.

Membrane cleaning results. The standard CIP procedure used for the INDECK/Hydranautics membrane consisted of recirculating a pH 2 solution through the membrane system followed by a backwash to clean the membrane module. A fresh acidic solution (pH 2) was then circulated through the system, and a second backwash was performed. A high pH caustic + hypochlorite solution was then prepared and circulated through the system, and a third backwash was performed. Generally, INDECK/Hydranautics targets a permeability of ≥ 12 gfd/psi after a CIP. Table 2.7 provides the cleaning results for the INDECK/ Hydranautics membrane, and descriptions of modified CIP procedures follow.

The CIPs performed after Runs 1, 2, and 3 followed the standard protocol. The success of these cleans was marginal. Caustic was eliminated from the cleaning procedure after Runs 4, 5, and 6, thereby lowering the pH during hypochlorite cleans. As a result, recovered permeabilities increased substantially. It is likely that while the caustic cleans helped remove organic fouling, the high pH of the solution was causing the precipitation of calcium carbonate or other inorganic foulants. Therefore, it will be important to either eliminate caustic from the CIP protocol or perform a third acid cleaning procedure following the caustic cleaning.

The CIP performed after Run 6 recovered 100 percent of the permeability from the start of the run, indicating that the 1.2 mg/L of polymer dosed to the membrane during the last 1.5 days of Run 6 did not irreversibly foul the membrane.

Table 2.7

CIP results – INDECK/Hydranautics (membrane permeability at reception* = 13.3 gfd/psi)

Run	Permeability at end of run (gfd/psi)	Permeability after CIP (gfd/psi)
1	11.3	10.7
2	8.6	11.6
3	6.4	8.6
4	5.3	**13.7**
5	13.4	**15.7**
6	6.3	**15.7**

* Specific normalized flux at start-up

Water quality summary. A summary of feed, permeate, and backwash water quality for INDECK/Hydranautics Run 1 is provided in Table 2.8. Similar tables for Runs 2 through 6 are located in Appendix A (Tables A.1 through A.5). The following feed water observations were made throughout pilot testing:

- Turbidities were high, with run medians ranging between 8 and 21 NTU.
- Particle counts were high, with run medians ranging between 9,700 and >18,000 counts/mL.
- pH values were typically greater than 10.
- Total chlorine ranged between 2.0 and 2.8 mg/L.
- Coliforms were not present; algal counts and HPCs were low.

The following trends in INDECK/Hydranautics membrane permeate were observed throughout pilot testing:

- Permeate turbidities were ≤ 0.06 NTU.
- Permeate particle counts were typically < 1 count/mL.
- Membrane filtration did not appreciably impact total alkalinity, total hardness, organic carbon concentrations, HAA_5 concentrations, coliform concentrations, or heterotrophic bacterial concentrations.
- Algae in the feed water were removed across the membrane.

Preliminary Design Criteria

Membrane system design criteria for developing membrane system layouts are provided in Table 2.9 for gravity operation (Scenario 1) and pumped feed water operation (Scenario 2). All criteria are based on operation at full capacity utilization of 223 mgd at 20°C (240 mgd at 23°C). Additional consideration was given to specific criteria to account for scaling up from pilot- to full-scale. Integrity test intervals were set at 24 hours as specified in pending regulations, and testing time was increased from 10 to 30 minutes to account for system draining and membrane rewetting. CIP solution volumes were increased from the pilot scale to account for dead volume in membrane racks, storage tanks, recirculation piping, and rinsing volumes.

Plant integration issues and capacity requirements were coupled with these criteria to develop conceptual layouts for the retrofit. These additional criteria and layouts are covered in the next chapter of this text.

Table 2.8
INDECK/Hydranautics water quality data – (Run 1: 9/5/02-10/5/02)

Parameter	Feed water				Permeate water				Backwash water			
	ave.	max.	min.	#*	ave.	max.	min.	#	ave.	max.	min.	#
Temperature (°C)	22.7	28.0	17.0	C†	-	-	-	0	-	-	-	0
pH	10.2	10.4	10.0	5	10.2	10.3	10.0	5	10.2	10.3	10.2	5
TSS (mg/L)	17	28	2	10	1	2	0	2	228	410	34	3
Total alkalinity (mg/L as $CaCO_3$)	34	40	28	4	37	37	37	2	-	-	-	0
Total hardness (mg/L as $CaCO_3$)	106	116	93	4	102	102	102	1	113	113	113	1
TOC (mg/L)	2.0	2.1	1.9	3	1.8	1.8	1.8	3	2.2	2.2	2.2	1
DOC‡ (mg/L)	2.3	2.5	2.1	3	2.7	3.5	2.0	2	3.1	3.1	3.1	1
UV_{254} absorption (cm^{-1})	0.043	0.045	0.040	3	0.049	0.059	0.042	3	-	-	-	0
SUVA (L/mg•m)**	1.9	-	-	3	-	-	-	0	-	-	-	0
Total chlorine (mg/L)	2.5	2.7	2.5	4	2.5	2.6	2.4	3	2.5	2.8	2.3	4
HPC (CFU/mL)	3.5	6	1	2	1	1	1	1	-	-	-	0
Total coliforms (P/A)§	A	-	-	4	A	-	-	4	-	-	-	0
Algae (#/100mL)	4	-	-	1	A	-	-	1	-	-	-	0
TTHMs (µg/L)	-	-	-	0	-	-	-	0	-	-	-	0
HAA_5 (µg/L)	13.7	13.7	13.7	1	16.3	17.8	15.0	3	18.5	21.8	15.2	2
[redacted]	% not exceeding			#	% not exceeding			#	ave.	max.	min.	#
	50	95	5		50	95	5					
Turbidity (NTU)	8.1	9.6	6.2	C	0.05	0.05	0.04	C	38	162	3	5
Total particle counts (>2µm; #/mL)	9,717	13,527	7,210	C	0.13	0.25	0.13	C	[redacted]			

* Number of samples
† Continuous measurements
‡ The elution of filter organic material may have caused DOC concentrations to exceed TOC concentrations.
§ Present/Absent
** Calculated value

Table 2.9
Sizing criteria* - INDECK/Hydranautics

System sizing	Gravity operation	Pumped feed water operation	Note
Flux	41	85	gfd at 20°C
Mode of operation	Deposition	Deposition	
Backwash interval	60 minutes	30 minutes	Minutes at 240 mgd
Backwash duration	70	70	Seconds
Backwash volume/module	63	63	Gallons
CIP interval	30	30	Days
CIP duration	8	8	Hours
CIP volume/module	40	40	Gallons/module
IT interval	24	24	Hours
IT duration	30	30	Minutes
Net flux	36.0	73.4	gfd at 20°C
Process residuals			
Backwash residuals			
Volume	17.9 mgd		7% of permeate production
TSS	14.3 times feed concentration		At 93% recovery
Hypochlorite	<3 mg/L		As Cl_2
Other dissolved constituents	Equal to permeate concentration		
Cleaning residuals			
Volume	<40 gallon/module		
TSS	<1 mg/L		Permeate concentration
Hypochlorite	Equal to permeate concentration plus 200 mg/L		
TDS	Permeate plus 200 mg/L		Max acid dose
pH	2-10		Assumes only acid or hypochlorite cleaning

* Based on 240 mgd of permeate production at 20°C and 30-day cleaning intervals

Zenon

System Description

The treatment steps in the Zenon ZW-1000 system are shown on the PFDs in Figures 2.12 and 2.13 for dead-end filtration mode and backwash mode, respectively. A brief description of operating modes and other critical functions is presented below.

Membrane module. The characteristics of the ZW-1000 membrane module used in this study are summarized in Table 2.10. The membrane element has 400 square feet (ft^2) of AMSA. According the manufacturer's specifications, the pH tolerance of the membrane in filtration is 5 to 9.5. However, the manufacturer indicated prior to the study that the feed water pH at the plant (> 10) was permissible. The membrane and system components are compatible with the oxidant concentrations in the feed water (2-3 mg/L total chlorine). Module materials also allow the membrane to be cleaned at low pHs (< 3) and with elevated hypochlorite concentrations during CIP events. The membrane operates in an O/I filtration mode, and is backwashed from I/O.

Full-scale membrane elements are mounted horizontally within a cassette (analogous to a module). Individual cassettes are connected together in a "stack" through a common connection to the permeate manifold. Cassettes are mounted in an open tank, below the water line. The number of cassettes per stack is based on site-specific factors including required capacity, flux, and basin depth. In most applications, permeate pumps draw a vacuum through the permeate piping to the inside of the membrane fiber. Backwash and recirculation pumps are used to provide energy for cleaning operations.

Table 2.10
Zenon membrane and module characteristics

Membrane type	UF
Model	ZW-1000
Module dimensions	27.2" x 4.1"x 26.9" (L x W x H)
Maximum flux	40 gfd
Maximum TMP	12 psi (vacuum)
Area per module (ft^2)	400
Flow direction	Outside to inside
Nominal pore size (Micron)	0.02
Membrane material	PVDF (hydrophilic and non-ionic)
pH range [operation (cleaning)]	5-9.5 (2-11)
Chlorine tolerance	1,000 mg/L (cleaning) 1,000,000 µg/L Hrs (Lifetime)

Source: Adapted from Zenon Environmental, Inc., 2003.

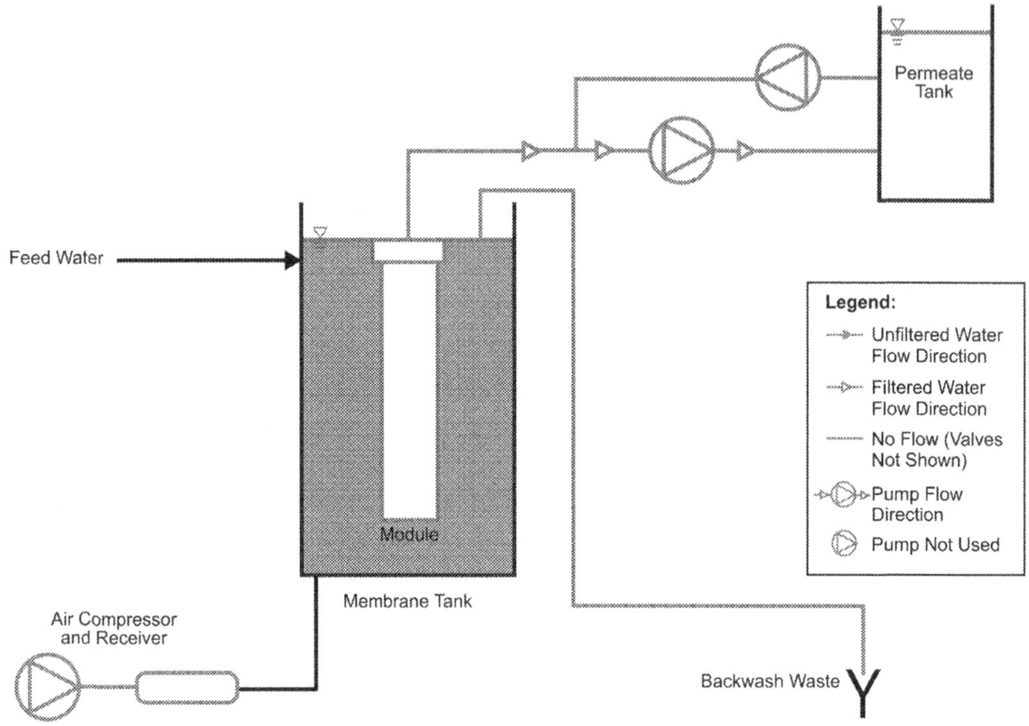

Figure 2.12 Zenon system PFD (filtration)

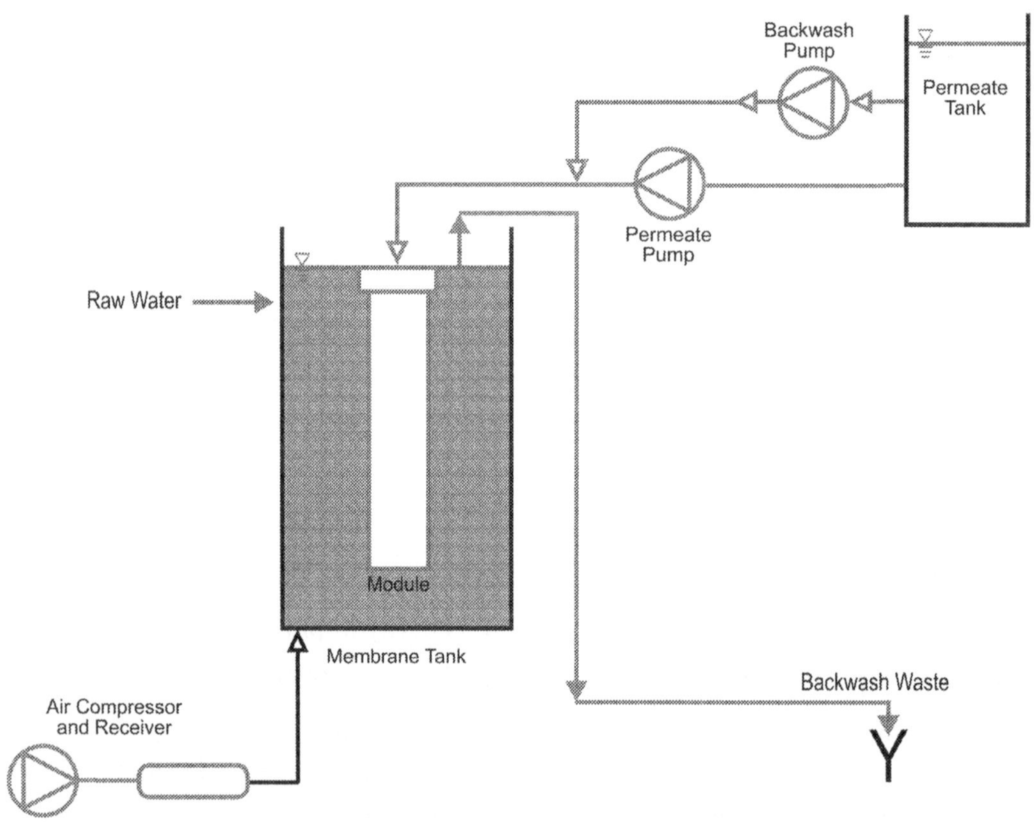

Figure 2.13 Zenon system PFD (backwash)

Filtration. During filtration a vacuum is drawn in the permeate piping and interior of the membrane fibers to pull the feed water across the membrane surface. Filtered water is then conveyed through a permeate pump and associated piping. Because the system operates with a vacuum, the maximum TMP is no more than atmospheric pressure (14.5 psi). However, equipment limitations and the potential for degassing restrict the maximum TMP from 10 to 12 psi (vacuum) in most applications, and lower at higher elevations. Compared to pressure-driven systems, flux rates are generally low. However, the fiber packing density (ft^2 membrane per ft^2 of footprint) is relatively high, resulting in similar required footprints.

Backpulse. Backpulses (backwashes) are performed at regular intervals (20 to 60 minutes) to remove solids from the membrane surface in order to maintain membrane permeability. Permeate water is forced through the membrane from I/O through the permeate piping at a rate of 0.7 to 2.7 times the filtration flow rate, depending on the application. Solids are removed from the tank through a backwash waste gullet. The reverse permeate flow is fed simultaneously with air scour, which assists in removing solids from the membrane surface and from the process tank. Periodically, solids accumulating on the bottom of the process tank are purged from the system through a bottom tank drain.

Permeate was backpulsed through the Zenon membrane at a flux such that the backpulse pressure did not exceed 10 psi (the terminal TMP of the membrane). Throughout the course of pilot testing, the backpulse flux (flow rate) and duration were set at 26.5 gfd (21.8 gpm) and 60 seconds, respectively. A 21.8 gpm backpulse flow rate is approximately one to two times the production flow rates used during pilot testing. Air scour was also used during each backpulse event at 1 standard cubic foot per minute (scfm). Per standard Zenon practice, no chemicals were added during backpulse events. One tank drain was performed daily, which was intended to further mitigate membrane fouling. During Runs 1 through 5, no chemicals were included in the tank drain procedure. During Run 6, 95 mg/L of hydrochloric acid was dosed to the process tank 15 minutes prior to the daily tank drain.

Clean-in-place. Backwashes help to reduce the TMP build-up but do not restore the full capacity of the membranes. The membrane must receive a chemical cleaning when the TMP reaches 10 to 12 psi. The chemical cleaning removes foulants from the membrane fibers that cannot be removed by backwashing alone. Membrane cleaning is typically performed by taking the system out of service and recirculating (in the direction of filtration – O/I) a low pH (pH = 2) solution, high pH (pH = 12) solution, and/or sodium hypochlorite (1,000 mg/L) as required.

Integrity testing. Membrane integrity is monitored with direct and indirect methods. Indirect methods include turbidity monitoring and particle counting. These parameters are measured with continuous monitoring devices, without the system being taken out of service. The system uses a PDT as the direct IT, which requires the membrane system to be taken out of service. During the test, the permeate side of the membrane fibers are pressurized, pressure in the feed side is monitored, and a decay rate is calculated. Pressure decay in excess of that associated with transmembrane air diffusion (0.75 psi/min) indicates a breach of membrane integrity. The IT requires approximately 2 minutes, but additional downtime results from draining the permeate side of the membrane prior to the test and rewetting the membrane following the test. Total system downtime required for the IT is approximately 10 to 20 minutes.

Three PDTs were performed over the course of the pilot study; two following installation of the membrane module and a third during Run 2C. Each test result met passing criteria as shown in Table 2.11.

Pilot Test Results

Hydraulic performance. Table 2.12 lists the production set points for the Zenon pilot plant during Runs 1 through 6. As was the case for the INDECK/Hydranautics membrane, the Zenon membrane operated using an ambient flux set point during Runs 1 and 2. For all subsequent runs, a normalized flux set point was utilized. Zenon maintained a recovery of 95 percent throughout pilot testing. A graphical summary of membrane hydraulic performance for all non-PAC runs is presented in Figure 2.14.

Figure 2.15 provides a graphical illustration of temperature and hydraulic performance trends for the Zenon system during Run 1. Similar graphs for Runs 2 through 6 are provided in Appendix A (Figures A.5 through A.11). A summary of run lengths, temperature trends, and hydraulic performance data for all six Zenon runs is provided in Table 2.13. Feed water temperatures were warm during the first month of testing, but dropped substantially during Runs 2 through 6.

With an average normalized flux of 28.3 gfd, Run 1 resulted in a final TMP of 6.5 psi, well below the terminal membrane TMP of 10.0 psi. However, when the average normalized flux increased to 36.9 gfd, the terminal TMP was reached after 12 days of operation. A run time of only 7 days was achieved when the average normalized flux subsequently decreased to 33.2 gfd for Run 2b. Although lower than the average normalized flux of the 30-day Run 1, the average normalized flux during Run 2c (26.4 gfd) resulted in an 18-day run time.

Table 2.11
Zenon integrity test results

Date	Criteria	Result
09/06/02	0.75 psi/min	Pass
09/06/02	0.75 psi/min	Pass
11/12/02	0.75 psi/min	Pass

Table 2.12
Production set points – Zenon

Run	Flux @ 20°C (gfd)	Recovery (%)	PAC	Polymer
1	28.3 (ave.)	95	N	N
2a	36.9 (ave.)	95	N	N
2b	33.2 (ave.)	95	N	N
2c	26.4 (ave.)	95	N	N
3	20.0	95	N	N
4	25.0	95	N	N
5	20.0	95	Y*	N
6	27.0	95	N	N

* Dosed at 5 mg/L

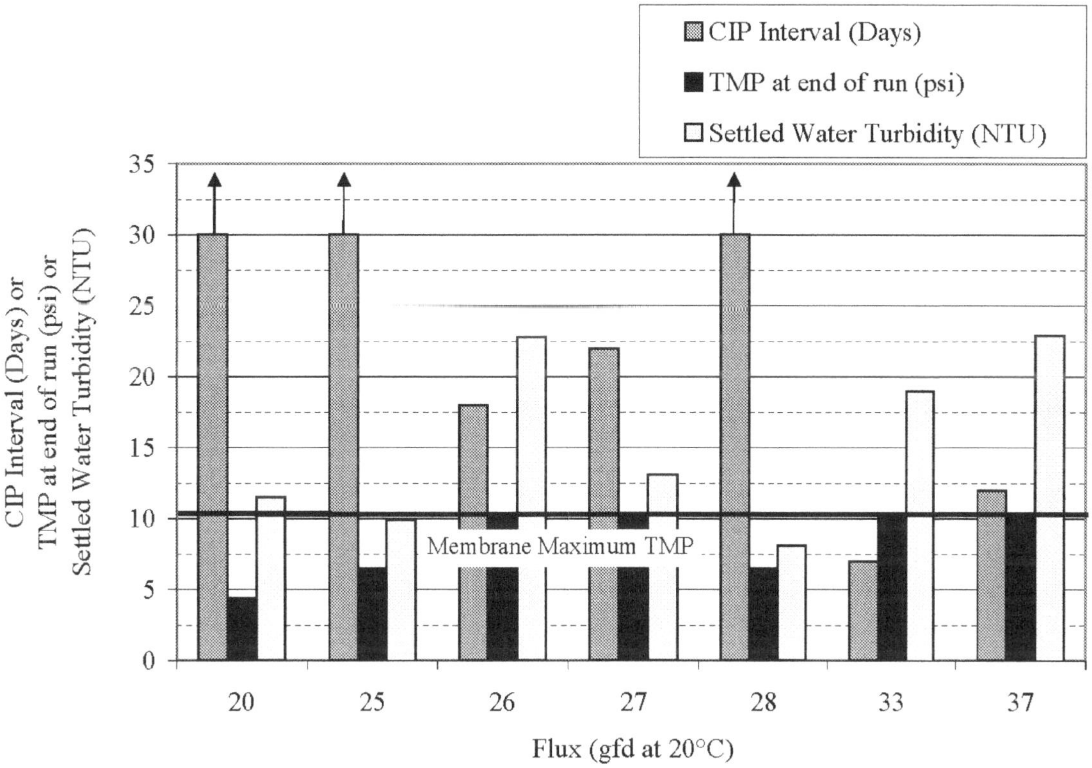

Figure 2.14 Hydraulic performance summary - Zenon

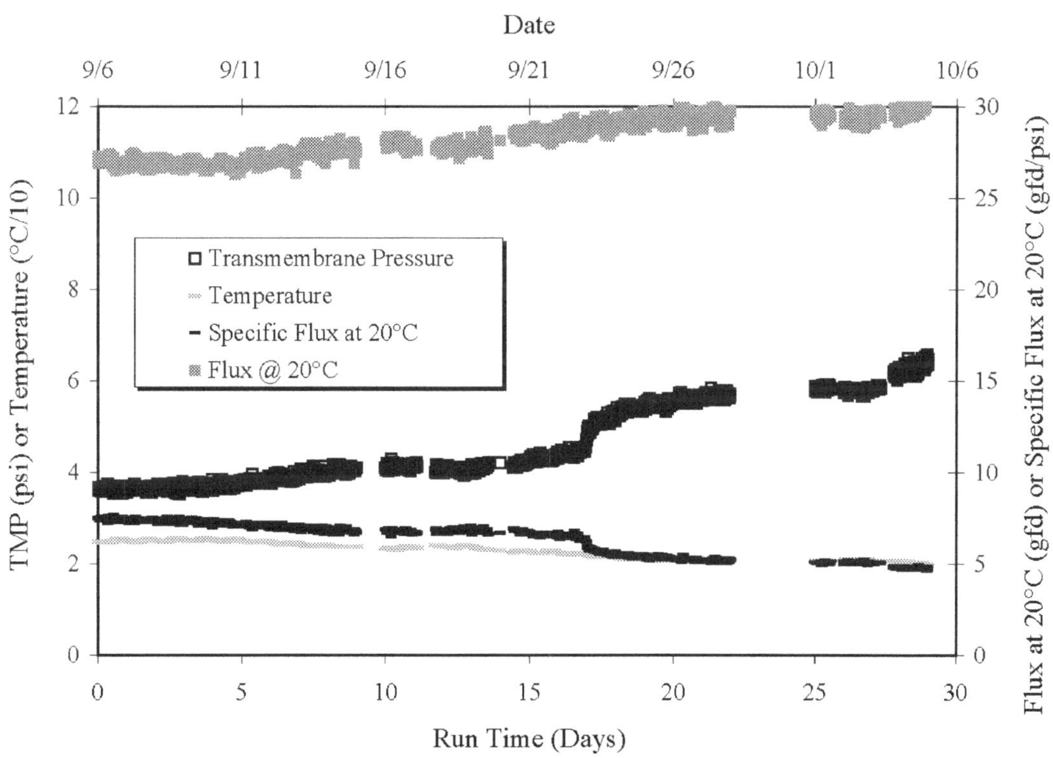

Figure 2.15 Run 1 Zenon hydraulic performance (operating conditions: average flux @ 20°C = 28.3 gfd, backwash interval = 25 minutes, recovery = 95%)

Table 2.13
Summary of Zenon hydraulic performance

Run #	Dates (2002 - 2003)	Length of run (days)	Temperature (range, average; °C)	Recovery (%)	Flux at 20°C (gfd)	Net Flux @ 20°C (gfd)	TMP (@ start of run, @end of run; psi)	Specific Flux @ 20°C (@ start of run, @ end of run; gfd/psi)
1	9/6-10/5	30	20-23, 23	95	28.3	27.2	3.6, 6.5	7.5, 4.7
2a	10/10-10/21	12	13-17, 15	95	36.9	35.4	6.8, 10.0	5.3, 4.2
2b	10/25-10/31	7	9-13, 10	95	33.2	32.2	6.3, 10.0	5.1, 3.3
2c	11/2-11/19	18	7-9, 9	95	26.4	25.8	5.0, 10.0	5.2, 2.6
3	11/21-12/20	30	4-8, 6	95	20.0	19.6	3.2, 4.5	6.4, 4.7
4	12/22-1/20	30	3-8, 4	95	25.0	24.5	5.1, 6.9	5.3, 3.1
5	1/24-2/22	30	3-7, 6	95	20.0	19.6	4.2, 4.6	5.1, 4.7
6	2/25-3/18	22	4-12, 8	95	27.0	26.2	5.0, 10.0	5.0, 2.7

While no clear relationship could be established between cleaning interval and flux, the impact of turbidity and flux together could explain the variability in cleaning intervals at similar flux rates. When median turbidities exceeded 10 NTU and flux exceeded 25 gfd at a 95 percent recovery, the membrane system did not meet the cleaning interval goal for the project. When flux was less than 25 gfd or when the median turbidity was less than 10 NTU, the membrane did meet the cleaning interval goal. Based on historical data for the years 2000 and 2001, average monthly turbidity exceeds 10 NTU 40 percent of the time. Because of the sensitivity of the membrane to feed water turbidity, a conservative design normalized flux of 20.0 gfd was chosen with a 95 percent recovery to ensure that 30-day run times would be achieved at full-scale.

As was observed with the INDECK/Hydranautics membrane, the 5 mg/L of PAC added to the feed water during Zenon Run 5 appeared to improve membrane performance. TMP increased by only 0.4 psi, and specific normalized flux values decreased by a total of 0.4 gfd/psi over the course of Run 5. These hydraulic results were better than those achieved during Run 3, which utilized identical production set points. There were no high polymer dose/response data gathered for the Zenon system due to the fact that Run 6 ended prior to the inception of the polymer dosing challenge test, having reached the terminal TMP at 22 days.

Data showing increased resolution around 3 backwash intervals during Run 3 are provided in Figure 2.16. As was observed in the INDECK/Hydranautics high resolution data, permeate particle counts tended to increase immediately following a backwash and decreased to near zero during the remainder of a given backwash interval. TMP fluctuated by approximately 0.4 psi throughout a given backwash interval.

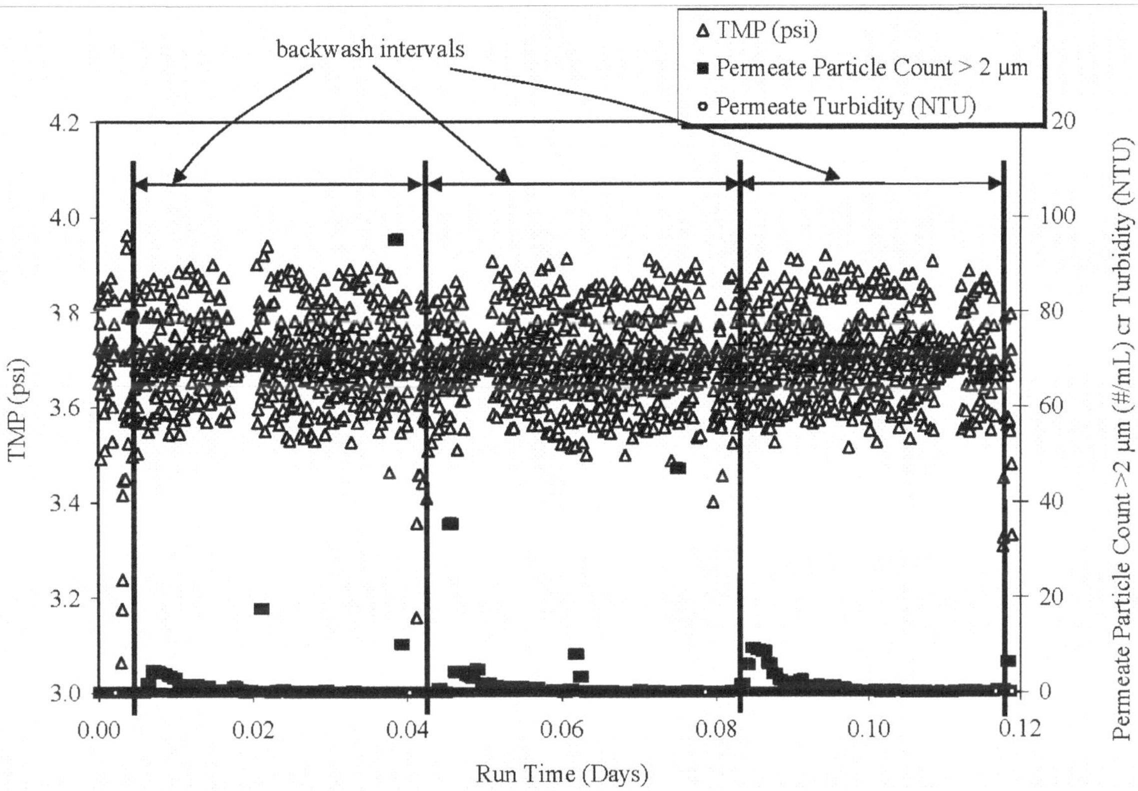

Figure 2.16 Run 3 Zenon hydraulic performance with zoom (operating conditions: flux @ 20°C = 20.0 gfd, backwash interval = 54 minutes, recovery = 95%)

Membrane cleaning results. The standard CIP procedure used for the Zenon membrane consisted of a citric acid backpulse and soak, followed by a hypochlorite backpulse and soak. A raw water and a clean water permeability were measured at the end of a given run and used as baselines for comparison. Permeability measurements were taken after the citric acid clean and again after the hypochlorite clean. Zenon targets a permeability minimum of 5.0 gfd/psi to determine the success of a CIP. Modifications to the standard CIP procedure can be made based on permeability results. Table 2.14 provides the cleaning results for the Zenon membrane, and descriptions of modified CIP procedures follow.

The CIP performed after Run 1 followed standard protocol. The hypochlorite step provided the largest improvement in permeability, indicating the occurrence of organic fouling during Run 1. Given these results, only the hypochlorite clean was performed after Run 2a. Due to the short run times achieved during Runs 2a and 2b, a more aggressive CIP was performed after Run 2b. After citric acid was backpulsed through the membrane, hydrochloric acid was added to the process tank to lower the pH of the soaking solution to 2.2. The membrane was allowed to soak for approximately 24 hours. The purpose of the enhanced acid cleaning procedure was to remove any inorganic fouling that was not affected by the standard citric acid procedure. This acid clean improved the permeability to 5.0 gfd/psi, and the subsequent hypochlorite clean provided an additional 0.6 gfd/psi. CIPs performed after Runs 2c, 3, 4, 5, and 6 also used the citric/hydrochloric acid procedure. In each case except for after Run 3, the acid clean was followed by a hypochlorite clean. The improvement in permeability after acid and hypochlorite cleans indicates that inorganic and organic fouling was occurring throughout pilot testing.

Table 2.14
CIP results – Zenon (membrane permeability at reception* = 7.5 gfd/psi)

	Membrane permeability (gfd/psi)			
Run #	At end of run	Clean water at end of run	After citric clean	After hypochlorite clean
1	5.3	5.5	**5.7**	**6.6**
2a	5.2	-	-	**5.9**
2b	3.8	4.1	**5.0**	**5.6**
2c	2.4	3.2	4.8	**6.8**
3	5.0	5.3	**6.1**	-
4	3.8	4.4	4.7	**5.3**
5	4.2	4.5	**5.3**	**5.7**
6	2.9	3.3	3.8	**5.7**

* Specific normalized flux at start-up

Water Quality Summary

A summary of feed, permeate, and backwash water quality for Zenon Run 1 is provided in Table 2.15. Similar tables for Runs 2 through 6 are located in Appendix A (Tables A.6 through A.12). Feed water quality trends described in the *INDECK/Hydranautics – Water Quality Performance* section also apply to the Zenon system.

The following trends in Zenon process streams were observed throughout pilot testing:

- Permeate turbidities were ≤ 0.04 NTU.
- Permeate particle counts were typically < 1 count/mL.
- Membrane filtration did not appreciably impact total alkalinity, total hardness, organic carbon concentrations, HAA_5 concentrations, coliform concentrations, or heterotrophic bacterial concentrations.
- Algae in the feed water were removed across the membrane.
- Backwash wastewater contained low HAA_5 concentrations.

Preliminary Design Criteria

Membrane system sizing criteria provided in Table 2.16 are based on full plant flow of 240 mgd. These criteria were developed using the pilot testing data and also incorporated full-scale considerations. CIP solution and tank drain volumes were estimated based on the pilot data; however, a review of the subsequent layouts revealed that these volumes were not directly scalable due to dead volumes in process tanks and pipe loops at the full-scale. The volumes in the following table reflect these considerations. IT intervals reflect proposed regulations, while IT durations include draining, testing, and rewetting requirements.

Table 2.15
Zenon water quality data – (Run 1: 9/6/02-10/5/02)

Parameter	Feed water				Permeate water				Backwash water			
	ave.	max.	min.	#*	ave.	max.	min.	#	ave.	max.	min.	#
Temperature (°C)	22.5	22.5	19.8	C†	-	-	-	0	-	-	-	0
pH	10.2	10.4	10.0	5	10.1	10.3	9.9	5	10.1	10.2	10.0	5
TSS‡ (mg/L)	17	28	2	10	8	15	1	2	8	21	0	3
Total alkalinity (mg/L as $CaCO_3$)	35	40	28	4	37	37	37	1	-	-	-	0
Total hardness (mg/L as $CaCO_3$)	106	117	93	4	107	114	101	3	103	103	103	1
TOC (mg/L)	2.0	2.1	1.9	3	1.9	2.0	1.9	2	3.2	3.2	3.2	1
DOC (mg/L)§	2.3	2.5	2.1	3	2.3	2.4	2.3	2	3.2	3.2	3.2	1
UV_{254} absorption (cm^{-1})	0.043	0.045	0.040	3	0.049	0.053	0.044	3	0.170	0.170	0.170	1
SUVA (L/mg•m)††	1.9	-	-	3	-	-	-	0	-	-	-	0
Total chlorine (mg/L)	2.5	2.7	2.5	3	2.6	5	1	3	-	-	-	0
HPC (CFU/mL)	3.5	6	1	2	3	5	1	2	-	-	-	0
Total coliforms (P/A)**	A	-	-	4	A	-	-	4	-	-	-	0
Algae (#/100mL)	4	-	-	1	A	-	-	1	-	-	-	0
TTHMs (µg/L)	-	-	-	0	-	-	-	0	-	-	-	0
HAA_5 (µg/L)	13.7	13.7	13.7	1	19.5	22.9	14.4	3	14.4	14.4	14.4	1
[redacted]	% not exceeding			#	% not exceeding			#	ave.	max.	min.	#
	50	95	5		50	95	5					
Turbidity (NTU)	8.2	10.5	6.4	C	0.04	0.04	0.04	C	-	-	-	0
Total particle counts (>2µm; #/mL)	9,717	13,527	7,210	C	0.2	2.2	0.0	C	[redacted]			

* Number of samples
† Continuous measurements
‡ TSS values may be high in spite of low turbidity and particle counts due to small sample volumes and analytical sensitivity.
§ The elution of filter organic material may have caused DOC concentrations to exceed TOC concentrations.
** Present/Absent
†† Calculated value

Table 2.16
Scenario 2 sizing criteria* - Zenon

	Scenario 2	Note
Flux	20	gfd at 20°C
Mode of operation	Deposition	
Backwash interval	45-50 minutes	At 240 mgd
Backwash duration	60	Seconds
Backwash volume/module	7.4	Gallons
Tank drain interval	24	Hours
Tank drain duration	20	Minutes (est.)
Tank Drain volume/module	50	Gallons (est.)
CIP interval	30	Days
CIP duration	8	Hours
CIP volume/module	50	Gallons (est.)
IT interval	24	Hours
IT duration	30	Minutes
Net flux	18.2	gfd
Process residuals		
Backwash/tank drain residuals		
Volume	12.6 mgd	5% of permeate production
TSS	20 times feed concentration	At 95% recovery
Total chlorine	<1 mg/L	As Cl_2
Other dissolved constituents	Equal to permeate concentration	
Cleaning residuals		
Volume	<40 gallon/module	
TSS	<1 mg/L	Permeate concentration
Hypochlorite	Equal to permeate concentration plus 200 mg/L	
TDS	Permeate plus 200 mg/L	Max acid dose
pH	2-10	Assumes only acid or hypochlorite cleaning

* Based on 240 mgd of permeate production and 30-day cleaning intervals

Westech/Polymem

System Description

The treatment steps in the Westech/Polymem system process are shown on the PFDs in Figures 2.17 and 2.18 for filtration and backwash mode, respectively. A brief description of operating modes and other critical functions is presented below.

Membrane module. The characteristics of the Westech/Polymem UF120 S3 membrane module used in this study are summarized in Table 2.17. The module contains 1,227 ft^2 of AMSA. According the manufacturer's specifications, the pH tolerance of the membrane during normal filtration is 2 to 12. The membrane and system components are compatible with oxidant concentrations in the feed water (2 to 3 mg/L total chlorine). Module materials also allow the membrane to be cleaned at low pHs (< 3) required to remove scale and with elevated hypochlorite concentrations during CIP events. The membrane operates in an O/I filtration mode and is backwashed from I/O.

Full-scale membrane modules are mounted vertically in a rack with a single feed and permeate connection. In most applications, the system uses feed, backwash, and recirculation pumps to provide energy for filtration, backwash, and cleaning operations.

Typically, backwash occurs every 20 to 60 minutes of membrane operation for duration of approximately 60 seconds. Water is delivered to the inside of the membrane fiber during a backwash at approximately 30 psi and forces accumulated solids out through the feed side (outside) of the membrane fiber. Backwash water is discharged from the system through the feed water piping. In some applications, backwash water is amended with hypochlorite (up to 5 mg/L) to mitigate membrane fouling.

Table 2.17
Westech/Polymem membrane module characteristics

Membrane type	UF
Model	UF 120S3
Module dimensions	H=36"; Diameter=12"
Maximum flux	N/A
Maximum TMP	21 psi
Area per module (ft^2)	1,227
Flow direction	Outside to inside
Nominal pore size (Micron)	0.08
Membrane material	Polysulfone
pH range	2-11
Chlorine tolerance	200 mg/L (cleaning) 150,000 mg/L Hrs (Lifetime)

Source: Adapted from Westech/Polymem, 2000.

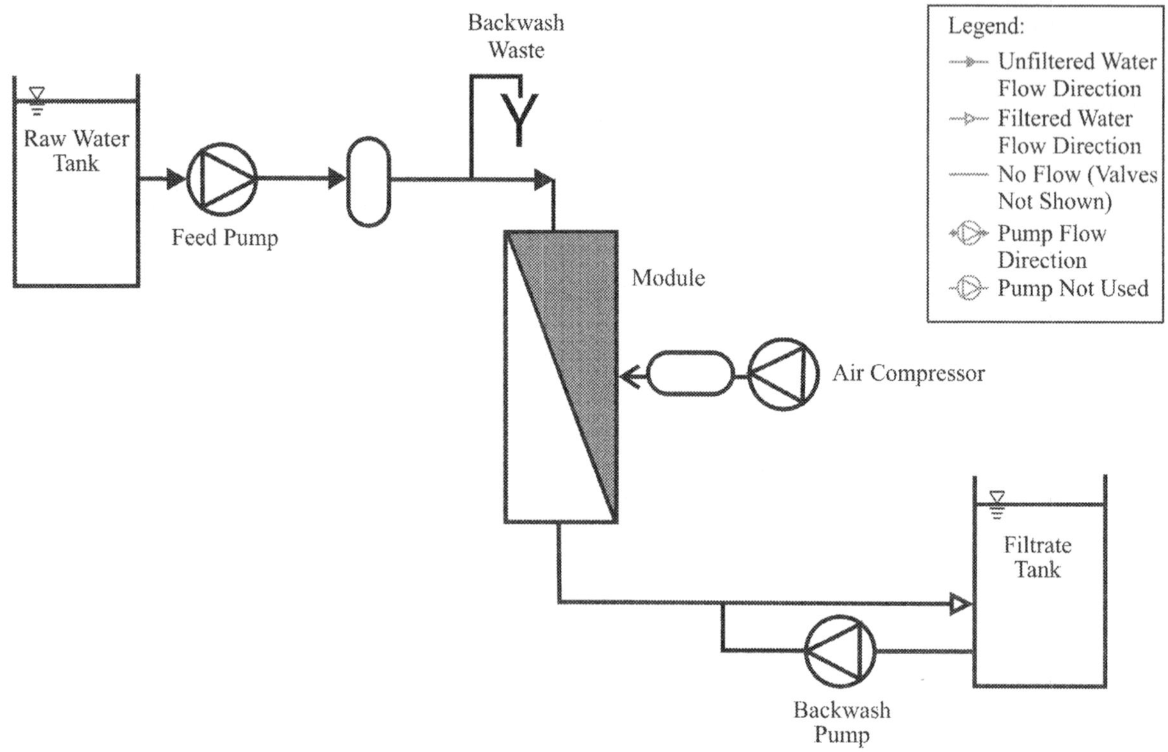

Figure 2.17 Membrane system PFD filtration

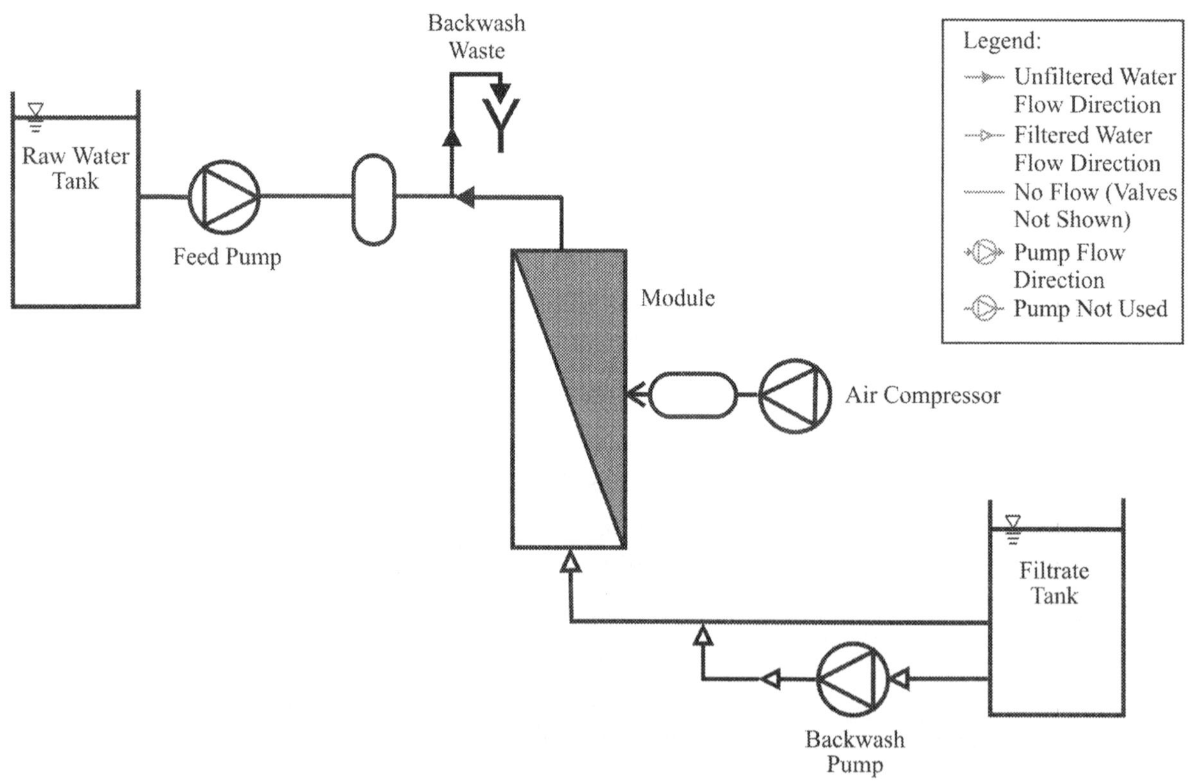

Figure 2.18 Membrane system PFD backwash

A 3-phase backwash protocol was used for the Westech/Polymem membrane during this study: (1) permeate, hypochlorite, and air were backwashed through the membrane for 20 seconds followed by a 5-second resting period, (2) permeate, hypochlorite, and air were again backwashed through the membrane for 20 seconds followed by a 5-second resting period, and (3) permeate and air were backwashed through the membrane for 20 seconds. Air and hypochlorite were always supplied at 2 scfm and 2 mg/L, respectively. Backwash pressure was set at 14.5 psi greater than the terminal production TMP of the membrane (~15 psi). This ~30 psi backwash pressure set point resulted in a backwash flow rate of approximately 100 gpm (117 gfd) throughout pilot testing.

Clean-in-place. Backwashes help to reduce the TMP build-up but do not restore the full capacity of the membranes. The membrane must receive a chemical cleaning when the terminal TMP for the run is met. The chemical cleaning removes foulants from the membrane fibers not removed by backwashing alone. Membrane cleaning is typically performed by taking the system out of service and recirculating (in the direction of filtration – O/I) at a low pH (pH = 2) solution, high pH (pH = 12) solution and/or sodium hypochlorite (200 mg/L) as required.

Integrity testing. Membrane integrity is monitored with direct and indirect methods. Indirect methods include turbidity monitoring and particle counting. These parameters are measured with continuous monitoring devices, without the system being taken out of service. The system uses a PDT as the direct IT, which requires the membrane system to be taken out of service. During the test, the feed side of the membrane fibers are pressurized, and the permeate side is left open to the atmosphere. Pressure in the feed side is monitored, and a decay rate is calculated. Pressure decay in excess of that associated with transmembrane air diffusion indicates a breach of membrane integrity. The IT requires approximately 10 minutes, and results in a total system downtime of 20 to 30 minutes due to the time required to drain the feed side of the membrane prior to the start of the test, and rewet the membrane following the test.

Four PDTs were performed over the course of the pilot study; one upon installation of the membrane module and subsequent testing following Runs 2, 3, and 4. One additional test was performed at day 5 of Run 4. With the exception of the initial test, fiber repairs were performed following each PDT. Summary of PDTs is provided below in Table 2.18.

Table 2.18
Westech/Polymem integrity test results

Date	Criteria	Result
10/15	0.73 psi/min	Pass
01/14	0.73 psi/min	Fail
01/28	0.73 psi/min	Fail
02/25	0.73 psi/min	Fail

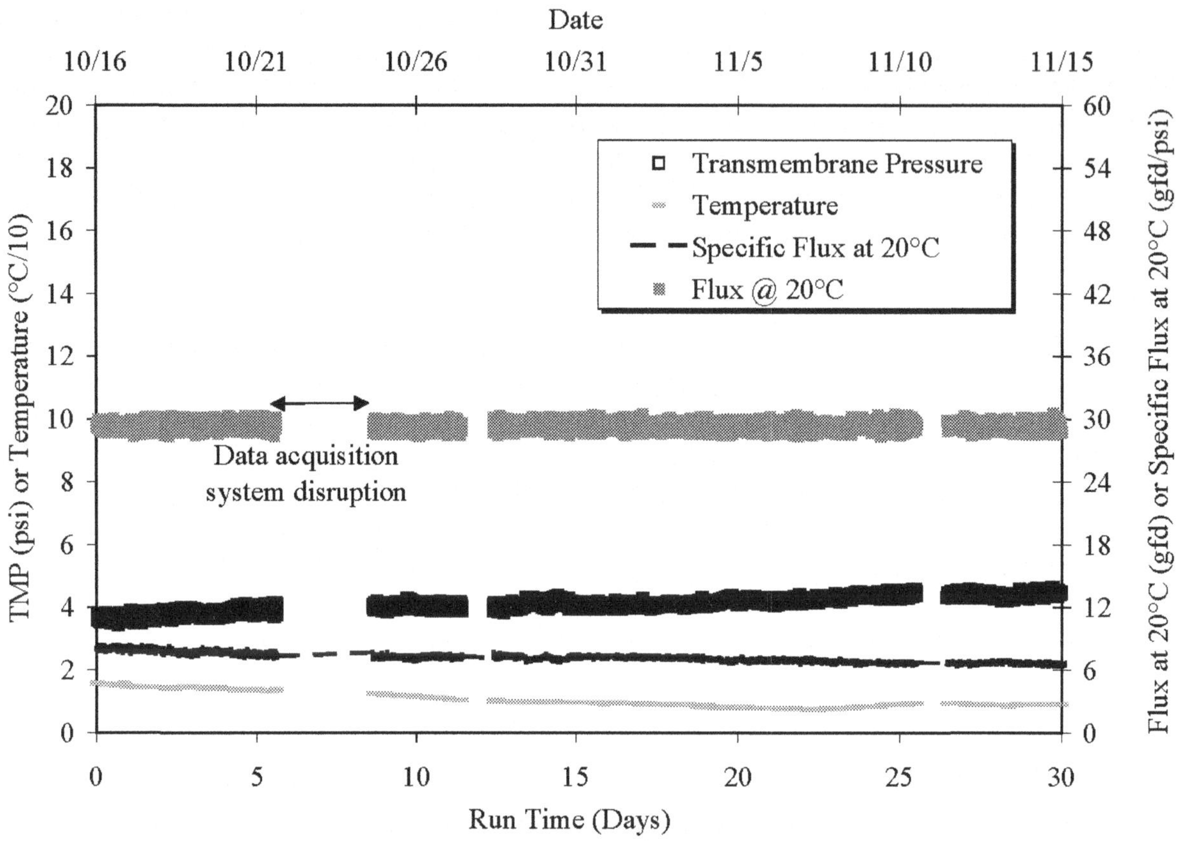

Figure 2.19 Run 1 Westech/Polymem hydraulic performance (operating conditions: flux @ 20°C = 29.7 gfd, backwash interval = 43 minutes, recovery = 90%)

Pilot Testing Results

Hydraulic performance. Table 2.19 lists the production set points for the Carollo pilot plant (Westech/Polymem membrane) during Runs 1 through 5.[2] Recovery was set at 90 percent during Runs 1 through 4 and was increased to 94 percent during Run 5. Figure 2.19 provides a graphical illustration of temperature and hydraulic performance trends for the Westech/Polymem system during Run 1. Similar graphs for Runs 2 through 4 are provided in Appendix A (Figures A.12 through A.14). A summary of run lengths, temperature trends, and hydraulic performance data for all five Westech/Polymem runs is provided in Table 2.20. A graphical summary of hydraulic performance is also provided for non-PAC runs in Figure 2.20.

[2] Note: The Carollo pilot plant with the Westech/Polymem membrane was started 1 month after the INDECK/Hydranautics and Zenon pilot plants. Therefore, Westech/Polymem Run 1 started simultaneously with Run 2 for INDECK/Hydranautics and Zenon.

Table 2.19
Production set points – Westech/Polymem

Run	Flux @ 20°C (gfd)	Recovery (%)	PAC	Polymer
1	29.7	90	N	N
2	35.3	90	N	N
3	41.2	90	N	N
4	41.2	90	Y*	N
5	56.0	94	N	Y†

*Dosed at 5 mg/L

† Dosed at 1.2 mg/L during final 3 days of Run 5

Table 2.20
Summary of Westech/Polymem hydraulic performance

Run #	Dates (2002 - 2003)	Length of run (days)	Temperature (range, average; °C)	Recovery (%)	Flux at 20°C (gfd)	Net Flux @ 20°C (gfd)	TMP (@ start of run, @ end of run; psi)	Specific flux @ 20°C (@ start of run, @ end of run; gfd/psi)
1	10/16 - 11/16	30	7-16, 11	90	29.5	28.9	3.6, 4.5	8.1, 6.6
2	11/16-12/16	30	4-9, 5	90	35.3	34.1	4.7, 6.2	7.4, 5.6
3	12/20-1/20	32	3-8, 4	90	41.2	39.8	4.7, 6.5	8.8, 6.4
4	1/23-2/26	32	3-7, 6	90	41.2	39.8	4.4, 5.8	9.4, 7.1
5	2/25-3/31	35	4-12, 8	94	56.0	54.7	6.8, 11.6	8.2, 4.8

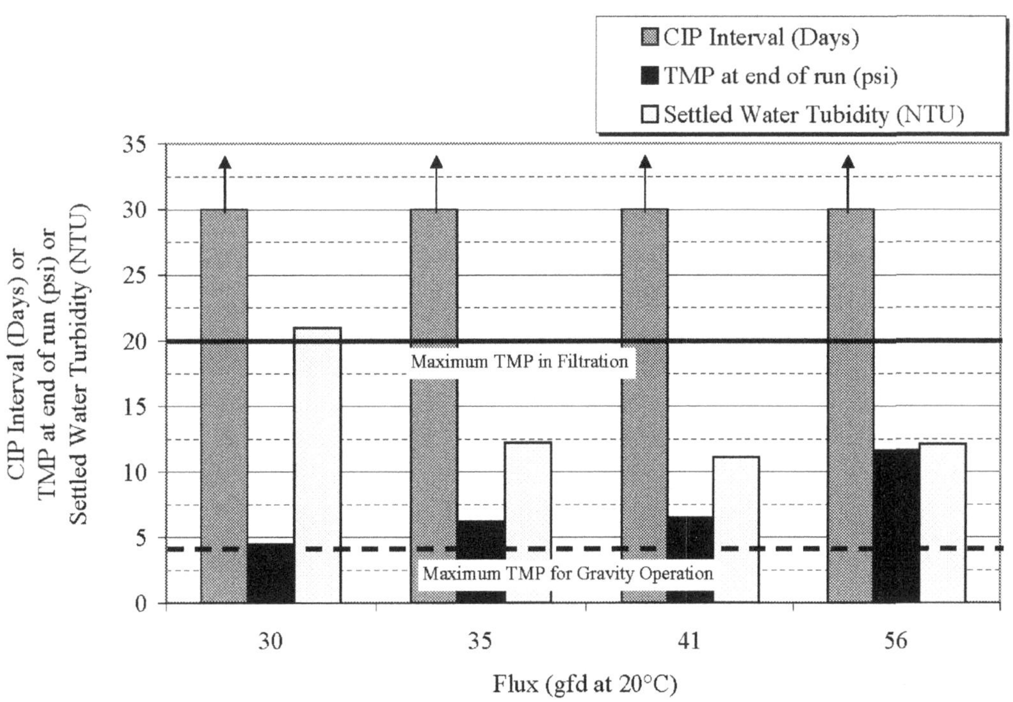

Figure 2.20 Hydraulic performance summary – Westech/Polymem

Run 1, which used a 29.7-gfd normalized flux set point, resulted in a TMP of 4.5 psi after 30 days of operation. During this time the median feed water turbidity was in the 80th percentile (21 NTU). The 30-day TMP was higher than the maximum available for gravity operation (4.3 psi); however, the 28-day TMP met the gravity operation criteria of 4.3 psi. Since each flux was considered for peak demand periods, it was determined that Run 1 flux would be appropriate for the gravity feed scenario.

The normalized flux set point was increased to 56.0 gfd for Run 5. After 32-days of run time, the TMP reached 11.6 psi and the feed pump shut down, as its capacity had been exceeded. Since the maximum normalized flux set point for the pilot plant had been reached, it was used as the design normalized flux for the pumped feed scenario.

As was observed with the INDECK/Hydranautics and Zenon membranes, the 5 mg/L of PAC added to the feed water during Westech/Polymem Run 4 appeared to improve membrane performance. TMP increased by only 1.4 psi and specific normalized flux values decreased by a total of 2.3 gfd/psi over the course of Run 4. These hydraulic results were better than those achieved during Run 3, which utilized identical production set points.

At the end of Run 6, 1.2 mg/L of low molecular weight, cationic polymer was dosed to the feed water continuously for three days. To ensure that the feed pump capacity would not be exceeded, normalized flux was reduced to 48 gfd at the inception of the polymer challenge test (Figure 2.21). Normalized flux was ramped up to 53 gfd over the course of the three day test. During this period, there was no appreciable change in TMP, even as normalized flux increased. Details related to the recovery of permeability after each run are provided in the *CIP Protocols and Results* section.

When data resolution was increased around three backwash intervals during the second Westech/Polymem run (Figure 2.22), TMP, particle count, and turbidity trends were observed to be similar to those noted during the INDECK/Hydranautics and Zenon backwash interval zoom. TMP increased and permeate particle counts decreased over the course of a given backwash. Permeate turbidity was not impacted by backwashing.

Membrane cleaning results. The standard CIP procedure used for the Westech/Polymem membrane consisted of recirculating a high-temperature (32 to 35°C) citric acid solution through the membrane system. This solution was maintained at a pH of 2 using small doses of hydrochloric acid. A high-temperature (32 to 35°C) hypochlorite solution was then recirculated through the membrane system. Multiple acid and hypochlorite cleans are often used to ensure oxidant/acid demand does not exceed oxidant/acid dose. Table 2.21 provides the cleaning results for the Westech/Polymem membrane, and descriptions of modified CIP procedures follow. Westech/Polymem requires that post-CIP permeability is ≥ 90 percent of the permeability at reception. (Initial permeability = 9.89 gfd/psi – cleaning goal = 8.8 gfd/psi).

Three acid cleans were performed at the end of Run 1, and recovery of permeability was marginal. Hypochlorite cleans were added to the CIP procedure after Run 2, and recovery of permeability increased to nearly 100 percent of the initial membrane permeability. Acid and hypochlorite cleans after Run 3 again provided good permeability recovery results. When a hypochlorite and caustic clean was performed after the hypochlorite cleans, permeability decreased. These results support the INDECK/Hydranautics CIP data that suggest the

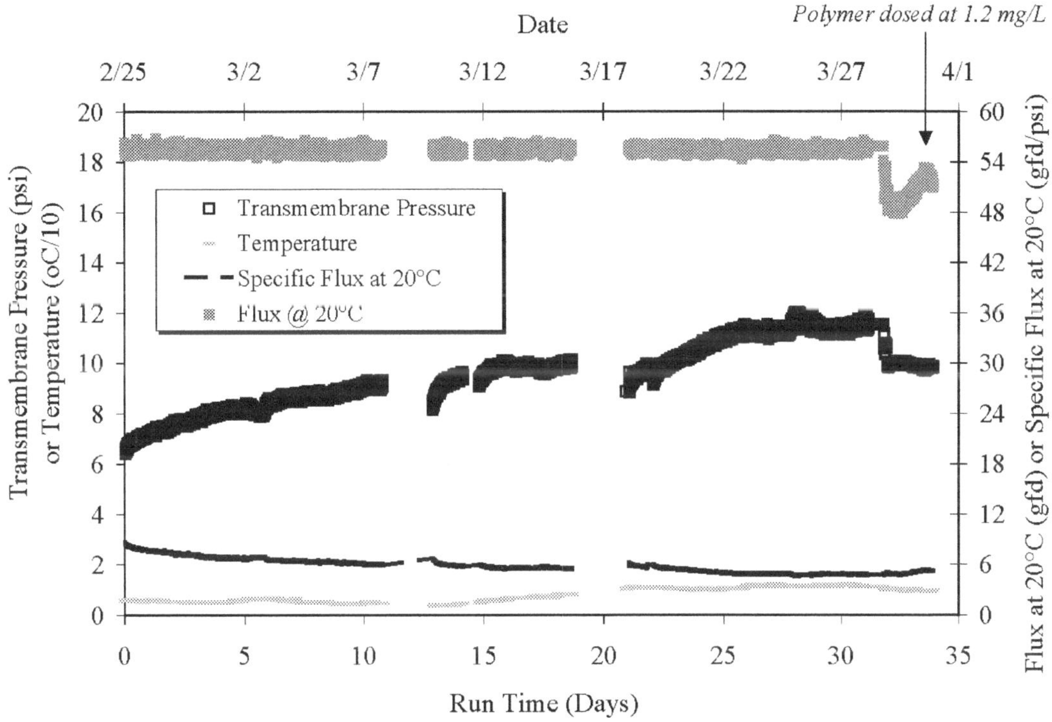

Figure 2.21 Run 5 Westech/Polymem hydraulic performance (operating conditions: flux @ 20°C = 56.0 gfd, backwash interval = 51 minutes, recovery = 94%)

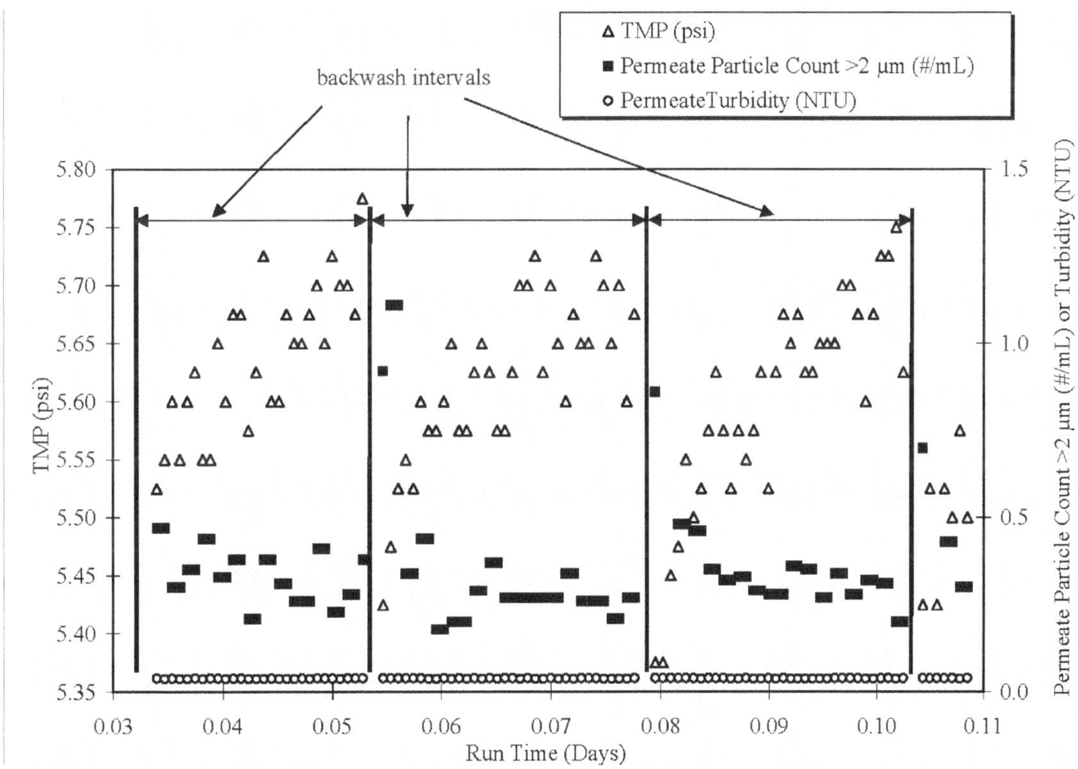

Figure 2.22 Run 2 Westech/Polymem hydraulic performance with zoom (operating conditions: flux @ 20°C = 35.3 gfd, backwash interval = 34 minutes, recovery = 90%)

Table 2.21

CIP results (gfd/psi) – Westech/Polymem (membrane permeability at reception* = 9.8 gfd/psi)

Run 1	After run	After acid clean 1	After acid clean 2	After acid clean 3	-	-	-	-	-	-
	6.6	6.9	7.5	7.8	-	-	-	-	-	-
Run 2	After run	After acid clean 1	After acid clean 2	After acid clean 3	After hypochlorite clean 1	After hypochlorite clean 2	After hypochlorite clean 3	-	-	-
	5.5	6.5	6.9	7.2	8.2	**8.9**	**9.5**	-	-	-
Run 3	After run	After acid clean 1	After acid clean 2	After hypochlorite clean 1	After hypochlorite clean 2	After hypochlorite clean 3	After hypochlorite clean 4	After hypochlorite + caustic clean	After hypochlorite clean 6	After acid clean 3
	5.6	6.9	7.3	7.7	**8.9**	**8.9**	**9.4**	**8.5**	**9.5**	**10.6**
Run 4	After run	After acid clean 1	After acid clean 2	After hypochlorite clean 1	After hypochlorite clean 2	After hypochlorite clean 3	-	-	-	-
	7.7	8.5	8.5	**9.2**	**9.3**	**9.7**	-	-	-	-
Run 5	After run	After acid clean 1	After acid clean 2	After hypochlorite clean 1	After hypochlorite clean 2	After acid clean 3	After acid clean 4	After hypochlorite clean 3	After acid clean 4	-
	5.3	5.9	6.1	6.0	6.2	6.6	6.8	6.9	7.4	-

* Calculated from a baseline TMP vs. normalized flux curve just before start-up

occurrence of precipitation during high pH cleans. Hypochlorite and acid cleans following the caustic clean resulted in greater than 100 percent recovery of the initial membrane permeability. These data suggest that the caustic clean may a useful organic removal step as long as a subsequent acid clean is performed to remove any inorganic fouling produced during the caustic cleaning.

The CIP performed after Run 5 recovered only 76 percent of the initial permeability from the start of the run. Though no caustic clean was performed to enhance the removal of organic fouling, the fact that five separate acid cleans and three separate hypochlorite cleans recovered only 76 percent of the permeability indicate that the 1.2 mg/L of polymer dosed to the membrane during the last three days of Run 6 irreversibly fouled the membrane.

Water quality summary. A summary of feed, permeate, and backwash water quality for Westech/Polymem Run 1 is provided in Table 2.22. Similar tables for Runs 2 through 6 are located in Appendix A (Tables A.13 through A.16). Feed water quality trends described in the INDECK/Hydranautics – Water Quality Performance section also apply to the Westech/ Polymem system.

The following trends in Westech/Polymem process streams were observed throughout pilot testing:

- Permeate turbidities were $\leq$ 0.04 NTU.
- Permeate particle counts were typically < 1 count/mL.
- Membrane filtration did not appreciably impact total alkalinity, total hardness, organic carbon concentrations, HAA_5 concentrations, coliform concentrations, or heterotrophic bacterial concentrations.
- Algae in the feed water were removed across the membrane.
- Backwash wastewater contained low HAA_5 concentrations.

Preliminary design criteria. Membrane system sizing criteria provided in Table 2.23 are based on full plant flow of 240 mgd. These criteria were developed based on pilot testing results and full-scale considerations. CIP solution and tank drain volumes factor dead volumes in pipe loops. IT intervals reflect proposed regulations, while IT durations include draining, testing, and rewetting requirements.

Membrane cleaning times and strategies were interpreted from the pilot data. It was apparent that the acid and hypochlorite cleans provided the greatest permeability recovery. However, multiple cleans were performed following each run to restore permeability, indicating a hypochlorite and acid demand. Cleaning times adopted for the Westech/Polymem system presented in Table 2.22 are more typical of the system, and are consistent with the INDECK/Hydranautics system.

Table 2.22

Westech/Polymem water quality data – (Run 1: 10/16/02-11/16/02)

	Feed water				Permeate water				Backwash water			
Parameter	ave.	max.	min.	#*	ave.	max.	min.	#	ave.	max.	min.	#
Temperature (^{o}C)	11.0	15.8	7.3	C†	-	-	-	0	-	-	-	0
pH	10.4	10.4	10.3	3	10.2	10.3	10.1	5	10.2	10.3	10.1	4
TSS (mg/L)	31	37	22	11	-	-	-	0	403	500	290	4
Total alkalinity (mg/L as $CaCO_3$)	48	56	43	4	32	32	32	1	36	36	36	1
Total hardness (mg/L as $CaCO_3$)	103	125	92	5	95	95	95	1	-	1	1	1
TOC (mg/L)	2.0	2.2	1.8	4	2.0	2.2	1.8	4	-	-	-	0
DOC (mg/L)	2.0	2.2	1.8	4	2.0	2.2	1.8	4	-	-	-	0
UV_{254} absorption (cm^{-1})	0.043	0.046	0.040	4	0.042	0.044	0.040	4	0.051	0.051	0.051	1
SUVA (L/mg•m)§	2.2	-	-	4	-	-	-	0	-	-	-	0
Total chlorine (mg/L)	2.5	2.8	2.0	5	-	-	-	0	3.1	3.8	2.4	4
HPC (CFU/mL)	2	6	0.5	4	8.8	17.0	0.5	2	-	-	-	0
Total coliforms (P/A)‡	A	-	-	4	A	-	-	4	-	-	-	0
Algae (#/100mL)	15	-	-	1	A	-	-	1	-	-	-	0
TTHMs (μg/L)	-	-	-	0	-	-	-	0	-	-	-	0
HAA_5 (μg/L)	11.1	12.0	9.7	4	11.3	12.9	10.0	6	-	-	-	0
[redacted]	% not exceeding			#	% not exceeding			#	ave.	max.	min.	#
	50	95	5		50	95	5					
Turbidity (NTU)	21.0	27.6	14.0	C	0.04	0.04	0.04	C	324	830	81	4
Total particle counts (>2μm; #/mL)	10,000	10,000	9,422	C	0.0	1.3	0.0	C	[redacted]			

* Number of samples
† Continuous measurements
‡ Present/Absent
§ Calculated value

Table 2.23
Sizing criteria* - Westech/Polymem

System sizing	Scenario 1	Scenario 2	Note
Flux	30	56	gfd at 20°C
Mode of operation	Deposition	Deposition	
Backwash interval	40 minutes	30 minutes	At 240 mgd
Backwash duration	70	70	Seconds
Backwash volume/module	100	100	Gallons
CIP interval	30	30	Days
CIP duration	18	18	Hours
CIP volume/module	40	40	Gallons/module
IT interval	24	24	Hours
IT duration	20	20	Minutes
Net flux	27.2	51.6	gfd
Process residuals			
Backwash residuals			
Volume	24.5 mgd		10% of permeate production
TSS	10 times feed concentration		At 90% recovery
Hypochlorite	<3 mg/L		As Cl_2
Other dissolved constituents	Equal to permeate concentration		
Cleaning residuals			
Volume	<40 gallon/module		
TSS	<1 mg/L		Permeate concentration
Hypochlorite	Equal to permeate concentration		
TDS	Permeate plus 200 mg/L		Max acid dose
pH	2-10		Assumes only acid or hypochlorite cleaning

* Based on 240 mgd of permeate production and 30-d cleaning intervals

CHAPTER 3
RETROFIT ASSESSMENT

Based on the results of the pilot test, two distinct scenarios for retrofit options were evaluated for the facility. The Scenario 1 layout was developed to maximize the use of existing filter infrastructure including feed channels, available filter driving head, backwash supply and waste lines, and filtered water piping. The existing filter backwash pumps would be used to backwash membranes. The Scenario 2 retrofit layouts were developed in a more typical fashion, resulting in a more autonomous unit filtration process. These layouts utilize feed or permeate pumps to provide membrane driving head and dedicated membrane backwashing pumps, as well as a range of new infrastructure for conveying water streams. Design criteria and layouts were developed for both scenarios and are discussed below.

The conceptual layouts were based on membrane-specific characteristics, capacity, and integration criteria developed for the plant. These criteria were developed to ensure that the conceptual layouts would meet capacity requirements, provide for membrane-specific operating requirements, and provide adequate space for operations and maintenance (O&M). This chapter is organized by scenario with discussions on how the critical functions of the membrane system were accomplished.

DESIGN CRITERIA

Capacity Requirements (Scenario 1 and 2)

Monthly water demand and temperature trends were used to determine the required membrane system capacity on an ambient temperature and temperature-normalized basis. Temperature normalization provides an understanding of the effect of water temperature on membrane system capacity for conditions under which the membranes were not tested. Temperature normalized flux rates can then be used directly for determining the required membrane surface area.

KCWSD provided daily filtered water volume and temperature data for the years 1999 through 2002. Monthly average production and maximum daily production are plotted in Figure 3.1. The average filtered water requirement series is representative of seasonal trends in the distribution system. Minimum demands occur from November to February. The maximum average water delivered over this period was 117 mgd, and maximum daily production was 134 mgd. The filtered water requirement peaks in July. The maximum monthly average filtered water produced in July was 162 mgd, and the peak daily production is 199 mgd.

Monthly water temperature at the plant varies significantly from winter to summer months. These temperatures range from 4°C (December through March) to a 25.5°C in July. The influence of temperature on membrane system capacity can be estimated by normalizing the demand data. Figure 3.1 shows a normalized demand, which is the result of temperature normalization of discrete combinations of minimum monthly temperatures and monthly maximum day filtered water requirements. The relative impact of temperature-related feed water viscosity changes and demand track each other closely resulting in a limited variation of normalized demand. During the months with the lowest minimum water temperatures (October through March), the maximum daily normalized demand varies between 182 and 198 mgd at 20°C. Due to warmer water temperatures, the normalized production decreases from April

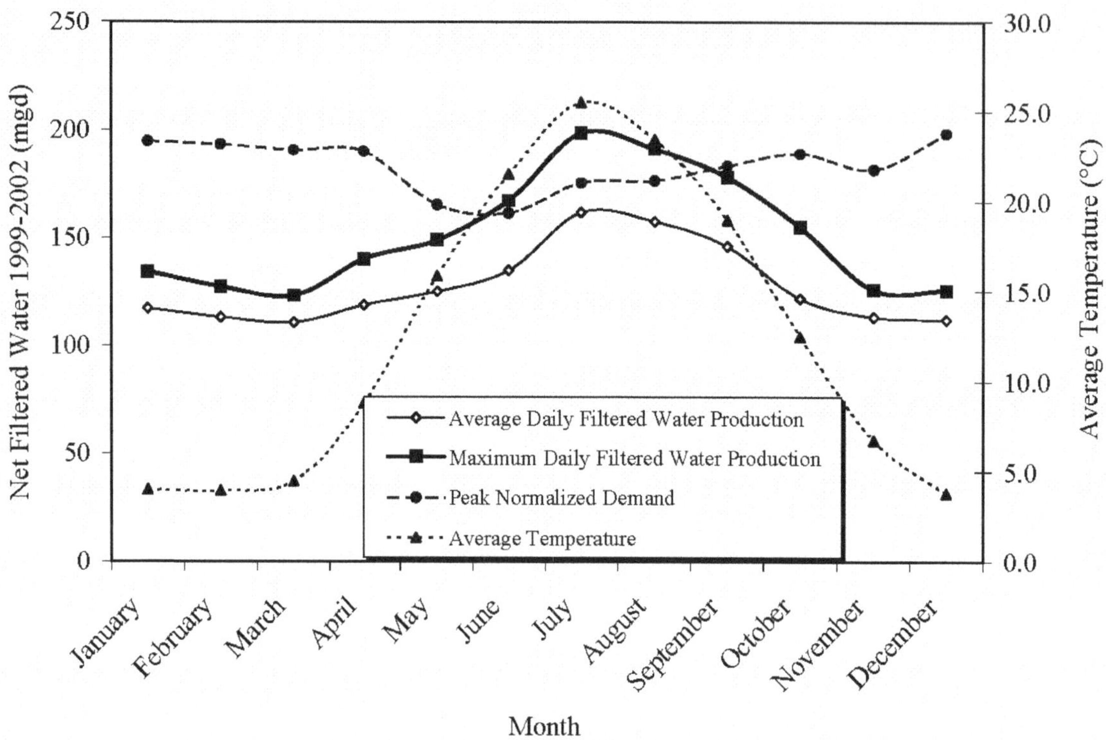

Figure 3.1 Temperature-normalized filtered water requirements

through September, despite the increase in the total filtered water requirement. During this period, the peak daily normalized filtered water requirement ranges from 161 mgd at 20°C (June) to 183 mgd at 20°C (September).

Based on this analysis, the minimum required membrane system capacity is 198 mgd at 20°C. However, in order to maintain the summer capacity of the plant, the membrane systems were sized to produce a minimum of 240 mgd at a temperature of 23°C (224 mgd at 20°C). Figure 3.2 shows the annual trend of estimated membrane system capacity at ambient temperatures based on the sizing criteria. In the summer months (June, July, and August), when the water temperature is in excess of 20°C, membrane system utilization ranges from 69 percent to 76 percent. Maximum membrane capacity utilization occurs during the months when water temperatures are less than 5°C. Capacity utilization during these months ranges from 79 percent to 89 percent.

Maximum System Headloss

The two retrofit scenarios have different criteria for selecting design flux rates. For Scenario 1, the membrane systems were sized to operate within the maximum achievable head between the settled water channel and the free water surface of the clearwell. As discussed in the previous chapter, operating conditions that resulted in TMPs of less than 10 feet (4.3 psi) after 30 days of operation were used in determining the required membrane surface area. For Scenario 2, the maximum flux rate that resulted in TMPs less than the maximum values specified by the membrane suppliers after 30 days of operation were adopted as the design flux rate.

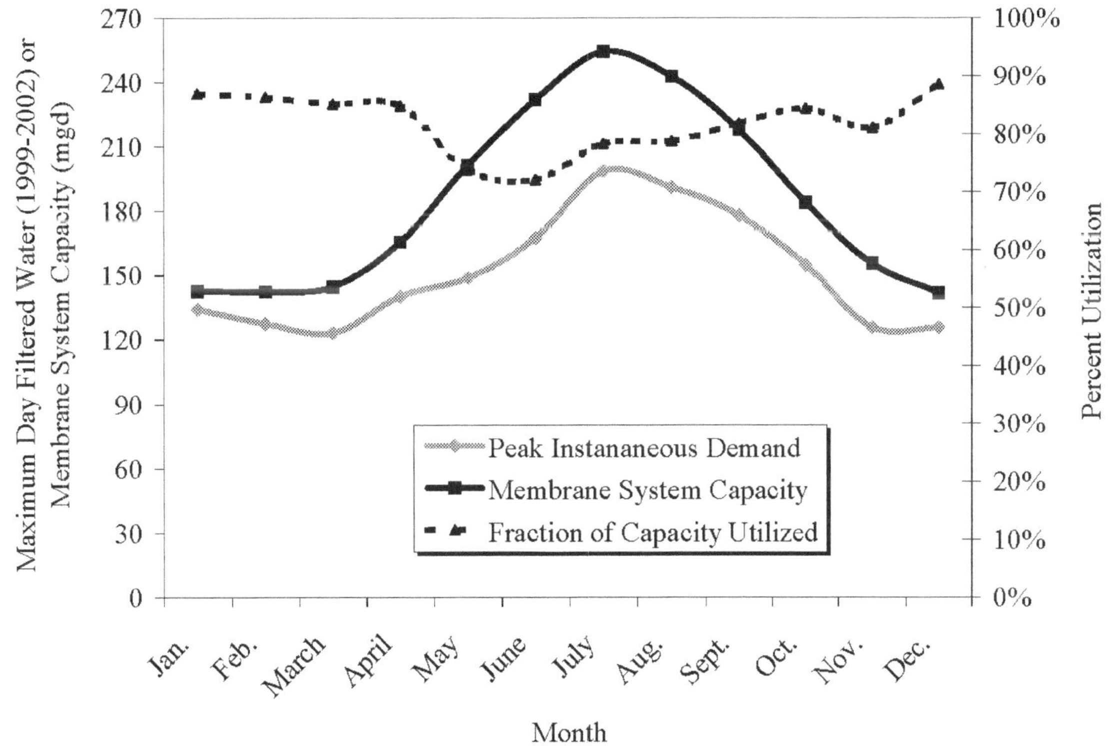

Figure 3.2 Membrane capacity utilization

The maximum achievable head was defined for this analysis as the elevation difference between free water surface elevation of the settled water channel and the clearwell. The estimated minimum water depth required in the high service pumping stations was the limiting factor for selecting the minimum clearwell depth. While reasonable modifications could be made in weir elevations to drop the clearwell free water surface elevation from 24 feet to 18 feet (assuming disinfection requirements are maintained), significant modifications to the existing high service suction wells or replacement of the pumps themselves would have to be made in order to accommodate a clearwell elevation lower than 18 feet.

To take advantage of the additional driving head resulting from a decreased clearwell elevation, the full pipe flow must be maintained in the permeate piping through its entrance into the clearwell. Additional piping would extend into the clearwell, with its exit pointing toward the clearwell floor. The weir that is used to maintain submergence of the existing filtered water piping would be lowered to approximately 16 feet to minimize the elevation behind the weir. The free water surface elevation over the weir would be 18 feet, resulting in a maximum available head of approximately 16 feet. To accomplish this, the clearwell would have to be operated at a maximum elevation of less than 18 feet, or 6 feet below the existing operating level. The resulting impacts on disinfection credits under the LT2ESWTR are not known and would have to be evaluated. It was assumed for the purposes of this analysis that the clearwell would be operated at a free water surface of 24 feet, providing 10 feet of driving head for membrane filtration.

Modifications described above would provide adequate additional head required to overcome membrane piping system headlosses.

Feed Water Screening

Feed water screening is generally required for each of the membrane systems evaluated in this study. To remove large particles from the membrane feed water, large particles can enter the open basins upstream of the filter gallery or flake off from the concrete channels and mortor-lined feed piping. These debris can cause damage to membrane surfaces and can increase the potential for lumen clogging. This can lead to permanent loss of membrane productivity and breaches in membrane system integrity.

Screening requirements vary by membrane supplier. In general, lumen clogging is not an issue with O/I membrane systems during filtration. Consequently, associated pre-screens have larger nominal separation sizes (500 μm to 2 mm). Screening in this range can be accomplished with a variety of strainers, including in-line self-backwashing screens, basket strainers, and band screens. However, selection of a screening device with minimum headloss is critical for Scenario 1 retrofits. Devices such as band screens could be installed upstream of the membrane system with less than 2 feet of head. The screens used for the O/I systems did not clog or require cleaning during the pilot study. For Scenario 2, losses across feed water screens do not impact feasibility, as feed water pumps can be sized to accommodate both screening and membrane system losses.

Lumen clogging in I/O systems can decrease membrane module life and greatly increase life cycle costs. Consequently, screens for these systems are generally sized between 130 to 200 μm. At one point during the pilot study, the feed water screen on the I/O membrane system clogged, and thus caused membrane system downtime. These screens have significantly higher headlosses and may preclude the use of I/O membranes for Scenario 1. However, in cases with extensive pretreatment upstream of the membrane system, membrane suppliers may provide a warranty a system without the use of screening, or with low-head, large-mesh screens. If screens are not used, implications on membrane supplier warrantees should be identified, and risks should be evaluated. Abbreviated warranties may render the approach infeasible due to membrane replacement costs. As with O/I systems in Scenario 2 losses across feed water screens do not impact feasibility for I/O systems, as feed water pumps can be sized to accommodate both screening and membrane system losses.

Membrane System Backwashing

The existing KCWSD granular media filters are backwashed using high service pumps. Water is diverted from the discharge of the high service pumps and into the individual filter galleries. A series of control valves between the high service pumps in each filter gallery control and regulate flow rates. Figure 3.3 shows the backwashing scheme at the facility. High service pumps 9 and 10 are used to backwash Gallery C. High service pumps 1 and 2 are used to backwash filters in Galleries A and B. Pump curves are provided in Figure 3.4. The backwash system is sized to accommodate flows up to 29,000 gpm. In the current backwash configuration, only one filter cell from either Gallery A or B may be backwashed at a time. A single filter cell can be backwashed in Filter Gallery C.

One potential limitation of using the current filter backwashing scheme for membrane systems is the existing condition of the backwash supply lines. The head required to backwash a membrane filtration system is significantly higher than that required to backwash the granular media filters. Head required for a granular media filter backwash is governed by friction losses

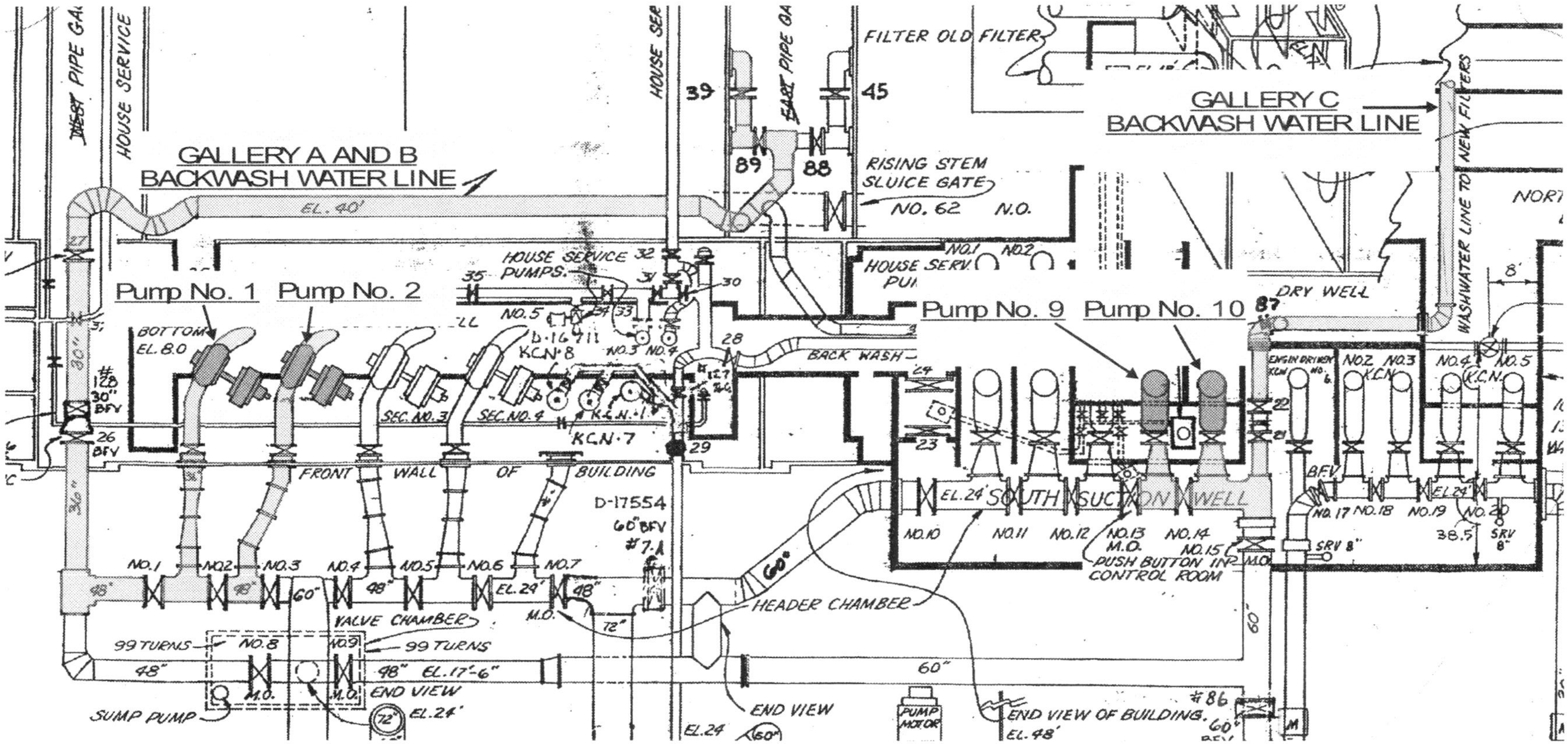

Source: Water Pollution Control Department, Kansas City, Missouri, 1979.

Figure 3.3 KCWSD filter backwash water routing

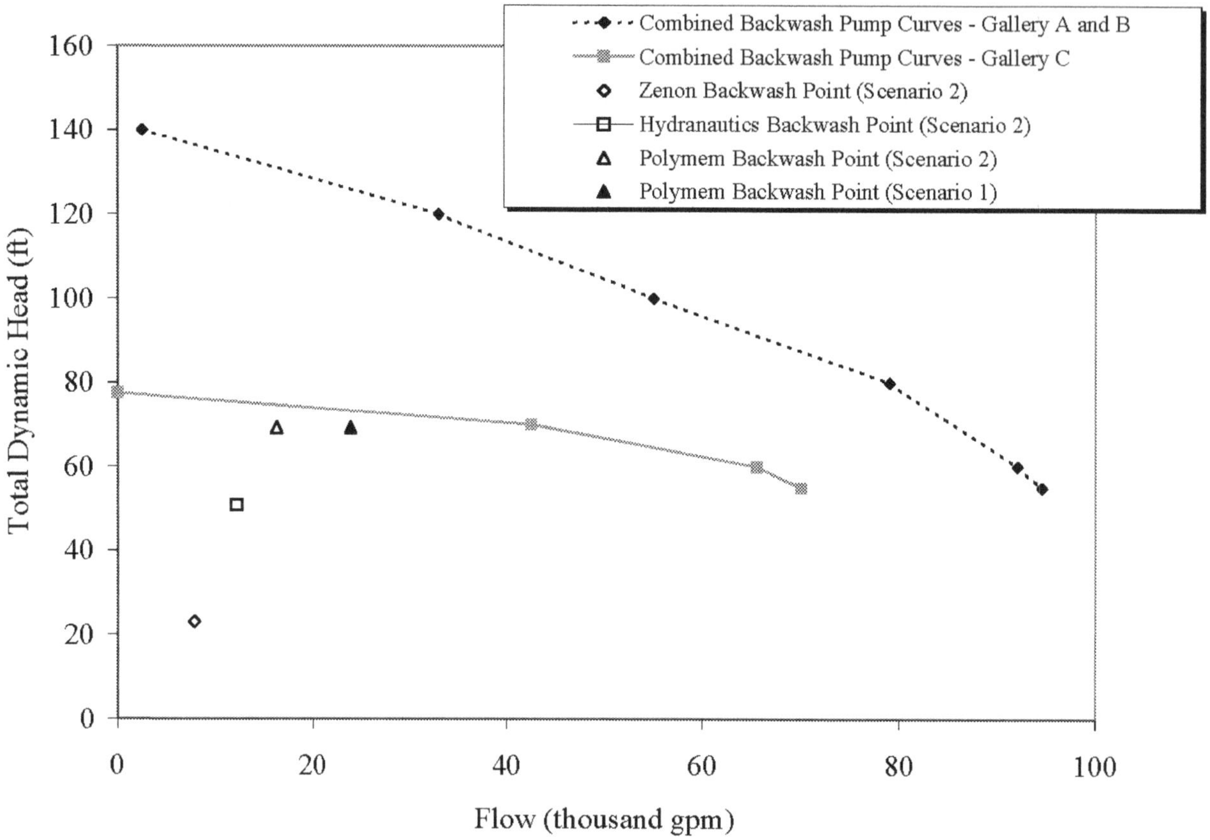

Figure 3.4 KCWSD backwash pump characteristics

and static head. For a membrane system, losses across the membrane surface govern the required backwash pump discharge head. The highest required backwash pressure for the three systems evaluated is approximately 30 psi. Control valves on the membrane racks would regulate the flow rate and modulate the on/off cycles required for the membrane backwash procedure.

While pump curves are adequate for the current condition, the backwash piping system in Kansas City would require some additional improvements or modifications to handle such pressures. Due to the age and condition of portions of the backwash supply piping and the nature of the pipe joints, maximum pressure in these lines is limited to 20 psi. Several options are available including repair of individual sections of pipe, lining the pipe, and reinforcing weaker joints. Scenario 1 layouts, which were developed to maximize integration of the membrane system within the existing infrastructure, assume that backwash pipe rehabilitation will be performed and that the existing high service pumps will be used for backwashing.

The use of dedicated pumps in each filter cell may allow KCWSD to forgo the task of backwash pipe rehabilitation. Booster pumps for each filter cell would allow the existing piping to continue to operate at its current pressures. New piping installed on the discharge side of the booster pump would accommodate the required membrane backwash pressure. This alternative was incorporated into all Scenario 2 layouts.

KCWSD staff indicated that pressure surges from the high service pumps occur, which can cause membrane fiber damage. In order to prevent damage from surges, air relief valves and devices designed to minimize rapid changes in velocity may be required at the plant. A surge study would be performed.

Backwash Supply Screening

Membrane backwash supplies generally do not require screening. However, because the plant has been in operation for nearly 80 years, there is potential for debris to accumulate in the high service pump clearwells that could cause lumen clogging on the permeate side of O/I membranes and potential fiber damage for I/O membranes. As a result, screening may need to be provided at the facility. This is not likely to cause hydraulic limitations, as adequate head can be provided by the high service pumps, at the flow rates required for backwashing. Screens would be provided with meshes near 500 μm for I/O membranes and 130 to 200 μm for O/I membranes.

Membrane System Operation and Maintenance

Other considerations were taken into account when developing the retrofit layouts, including system accessibility and operational convenience. The layouts were designed to leave sufficient space for accommodating CIP equipment, including recirculation pumps, day tanks, and equalization tanks. Skid arrangements and layouts within each individual filter cell were sized to provide minimum clearances for accessing individual membrane modules for removal and/or fiber repair.

SCENARIO 1: LAYOUTS

Layouts for Scenario 1 are described in this section and are divided by membrane supplier. Table 3.1 provides a comparison of the Scenario 1 equivalent hydraulic surface loading rates (SLRs) for INDECK/Hydranautics and Westech/Polymem layouts. Since one of the primary criteria of the Scenario 1 retrofit is that no additional structures are required, only layouts that resulted in a net SLR greater than the existing SLR were considered feasible. Figure 3.5 is a representation of the hydraulic grade line (filtration) for Scenario 1.

Table 3.1
Summary of equivalent filter surface loading rates

System	Scenario flux (gfd at 20°C)	Equivalent SLR (gpm/ft^2)
Existing filters	N/A	3.0
INDECK/Hydranautics	50	2.3
Westech/Polymem	30	3.5

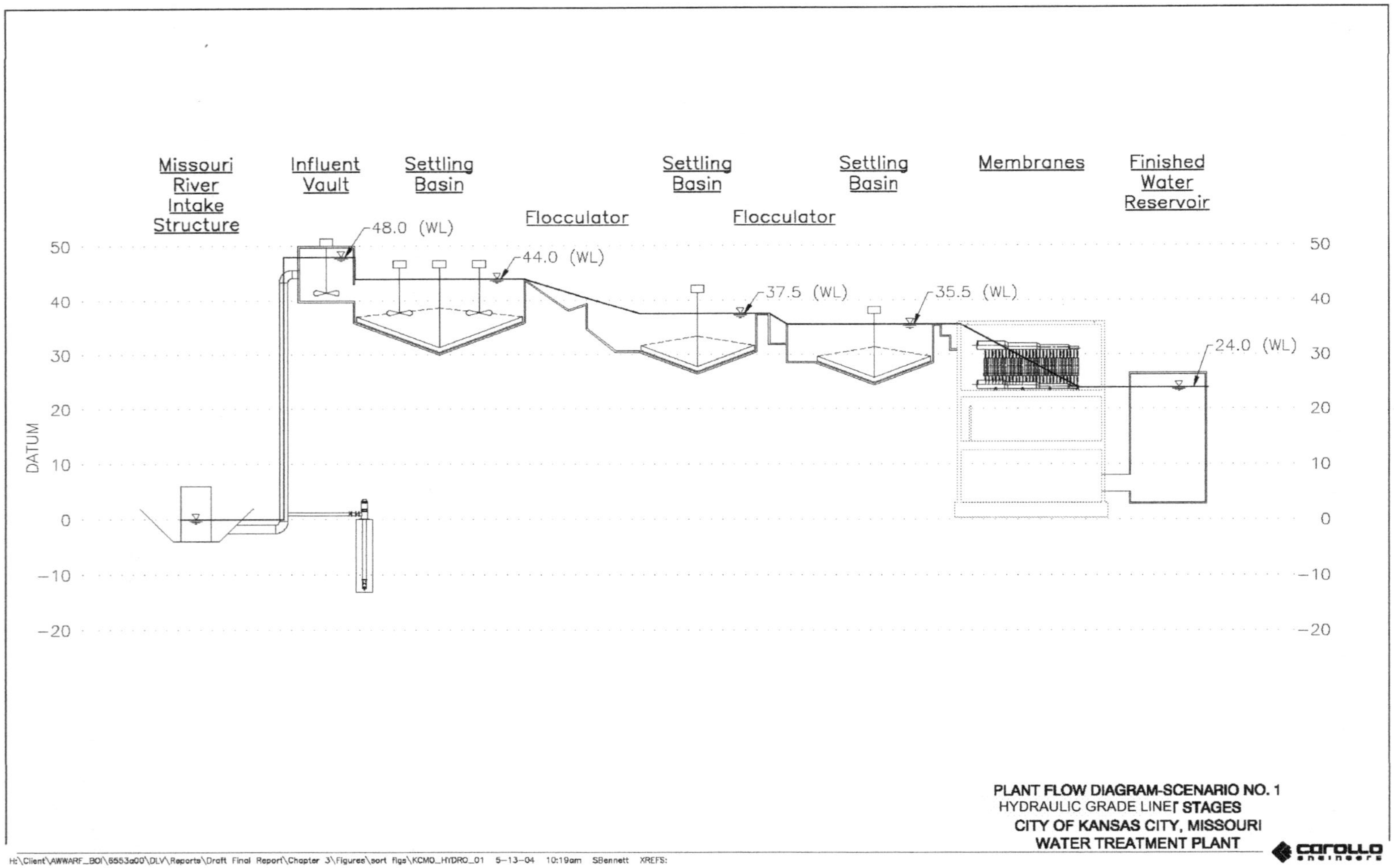

Figure 3.5 Scenario 1 hydraulic grade line

INDECK/Hydranautics

A layout was developed for the INDECK/Hydranautics membrane system that maximized the capacity of the filter gallery while maintaining the design criteria listed above. Based on the design operating parameters that met the criteria for Scenario 1 and conceptual layouts developed for Scenario 2, the projected total maximum daily production of the plant would be ≤ 164 mgd at 20°C. A flux of 72 gfd at 20°C would be required to meet the capacity requirement of this configuration (> 240 mgd), or 44 percent higher than the flux demonstrated in the pilot study.

In order to use this configuration at the flux rates demonstrated in the pilot study, a dedicated membrane building would have to be built to provide additional footprint for the membrane system (23,000 ft^2 minimum). Supporting infrastructure for water and power would also have to be installed. Because this system could not be used at KCWSD without substantial investment in dedicated infrastructure, it was not evaluated further for Scenario 1.

Similar facilities with more available driving head could feasibly utilize this membrane. With similar rack configurations and feed water quality, driving heads of 10 to 15 psi (23 to 35 feet of water) would result in an equivalent hydraulic SLR of greater than 3.0 gallons per minute per square foot (gpm/ft^2).

Westech/Polymem

Conceptual layouts for the Westech/Polymem system are provided in Figures 3.6 and 3.7 for Filter Gallery C. These layouts assume operation within the existing hydraulic grade line with a maximum flux of 30 gfd and a recovery of 90 percent, resulting in a net capacity of 240 mgd at 23°C. Based on projections from Westech/Polymem, the membrane system will require approximately 7,200 UF 120S3 membrane modules, each with an active membrane surface area of 114 m^2. Ten cells from each gallery (30 total) would be utilized for housing membrane racks. The two forward cells would be used for storing CIP tanks, chemicals, equalization basins, recirculation pumps, air compressors, and dosing pumps. Each filter cell would consist of 240 membrane modules. A minimum clearance of 3 feet is provided between the modules and the filter cell walls and approximately 3 feet is provided between rows of modules.

An elevation view of this layout provides some additional detail on the infrastructure reuse, specifically piping interconnections. Connections from the settled water channel to the membrane skid is accomplished using existing and new interconnecting piping to bring water to the membrane rack. Permeate water is then routed to a rate-of-flow controller and through the existing penetration to the clearwell. Flow control and measurement are accomplished with a pipe arrangement that accommodates two permeate water tie-ins, one from each side of the filter. Backwash waste is routed through the upper filter gullet through existing piping to the washwater sewer. CIP solutions are circulated through a pipe loop from one of the forward cells that contains the recirculation pump and chemical storage tanks (not shown). Spent CIP solution and rinse water is stored in a portion of the filter gallery for equalization and neutralization. Neutralized water is then disposed of in the existing washwater sewer.

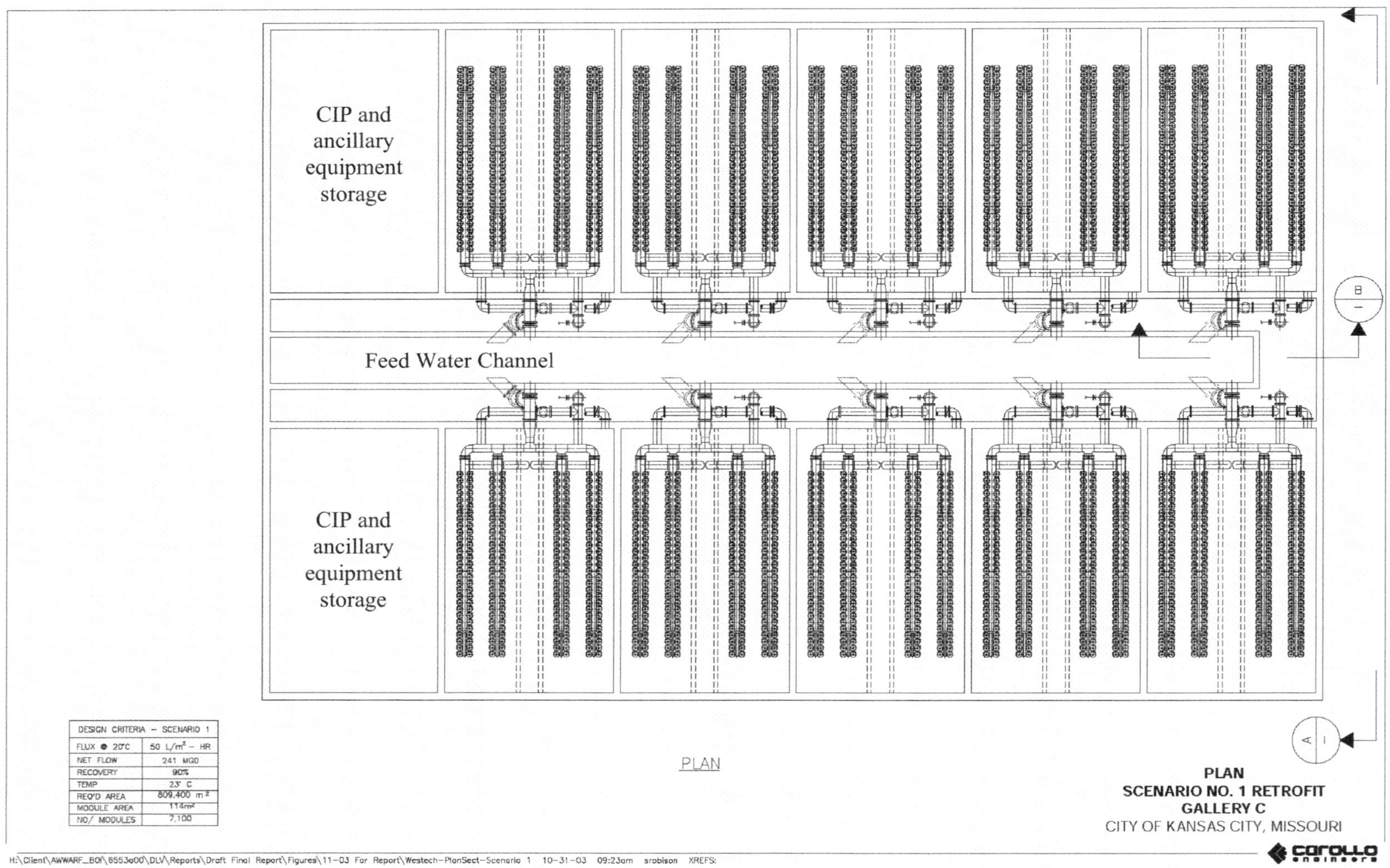

Figure 3.6 Westech/Polymem system Scenario 1 layout

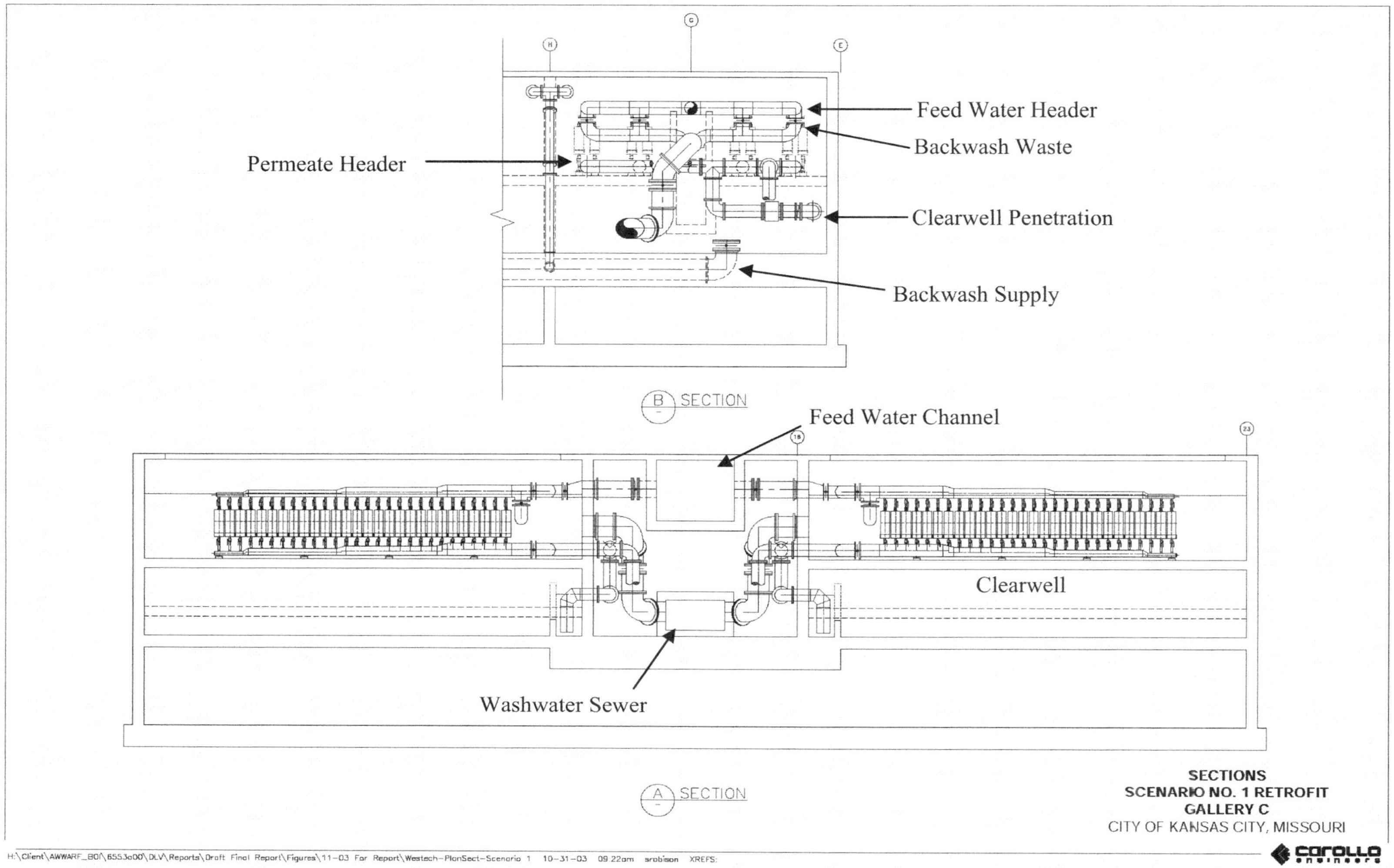

Figure 3.7 Westech/Polymem system Scenario 1 cross sections

Backwashing may be accomplished through the use of control valves and the existing plant backwash facilities. A single backwash event requires water from the high service pump for approximately 2 minutes. Assuming that 10 cells in each filter gallery are retrofitted with the membranes, a total of 20 individual filter cells will be in operation in Galleries A and B, thus requiring a total of 40 minutes to backwash all of the racks. For backwashing, Filter Gallery C (10 cells) is piped independent of Galleries A and B, and uses two different high service pumps. The total required duration of a backwash for Gallery C is 20 minutes. Based on pilot testing, backwash intervals are 40 to 50 minutes to maintain the minimum recovery. If this were not the case, membrane racks would be backwashed at longer intervals or be taken out of service until backwash supply was available. This could limit the membrane system capacity and/or decrease the cleaning interval. In this case, the backwashing approach is feasible and has been assumed for the cost evaluation.

SCENARIO 2: LAYOUTS

The concept for Scenario 2 layouts were based on the results of the pilot testing. The pilot testing indicated that fouling rates were much less than anticipated even at flux rates exceeding the recommended maximums indicated by the membrane suppliers. Therefore, the design criteria at the highest tested flux rates (which exceeded recovery and fouling criteria for the project) were used to size the configurations evaluated in Scenario 2. In Scenario 2, additional feed pressure is provided with feed water pumps, which allows the membranes to operate at higher flux rates. The higher flux rates resulted in a smaller required membrane area and produced adequate SLRs for each system tested (Table 3.2). A graphical representation of the Scenario 2 hydraulic grade line (filtration) for pressure driven systems is provided in Figure 3.8.

The flux and recovery criteria for Scenario 2 were given previously in Chapter 2. Other design criteria remain the same for spacing within the filter gallery, location of CIP, and other ancillary equipment within the filter boxes.

INDECK/Hydranautics

Typical plan and section drawings for the INDECK/Hydranautics system are provided in Figures 3.9 and 3.10, respectively. This system is sized to produce 240 mgd at 23°C and 156 mgd at 5°C with a 93 percent feed water recovery and a flux of 85 gfd at 20°C. A total of 6,336 modules are included in the layout. Because feed pumps are used, racks are located on top of the existing filter boxes, facilitating access for maintenance and repair activities.

Table 3.2
Comparison of operating conditions and equivalent surface loading rates

System	Scenario flux (gfd at 20°C)	Equivalent SLR (gpm/ft^2)*
INDECK/Hydranautics	85	3.4
Westech/Polymem	56	4.6
Zenon	20	7.4

* SLR presented as average rates at peak capacity in portions of the gallery with membrane racks.

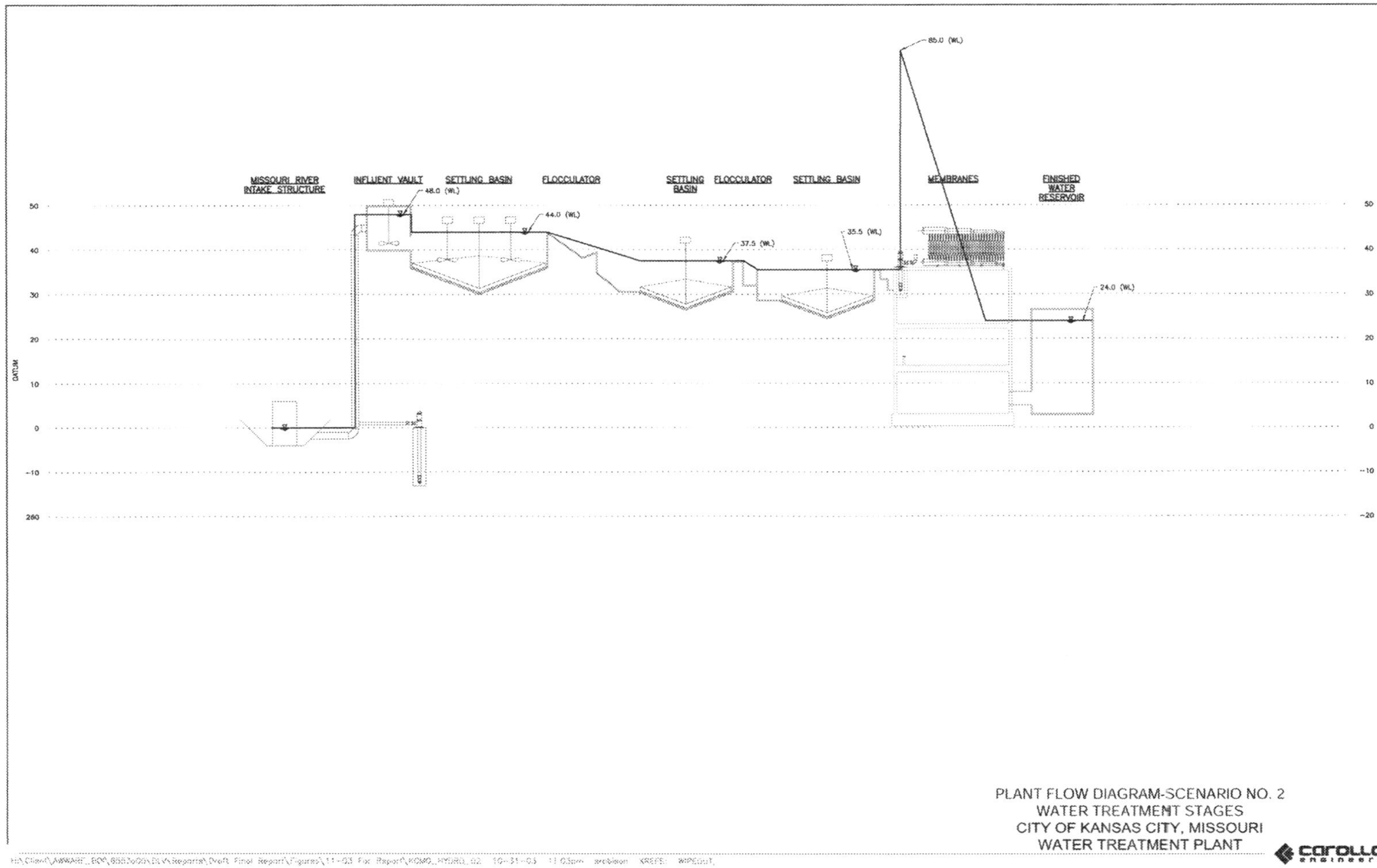

Figure 3.8 Scenario 2 hydraulic grade line – pressure-driven system

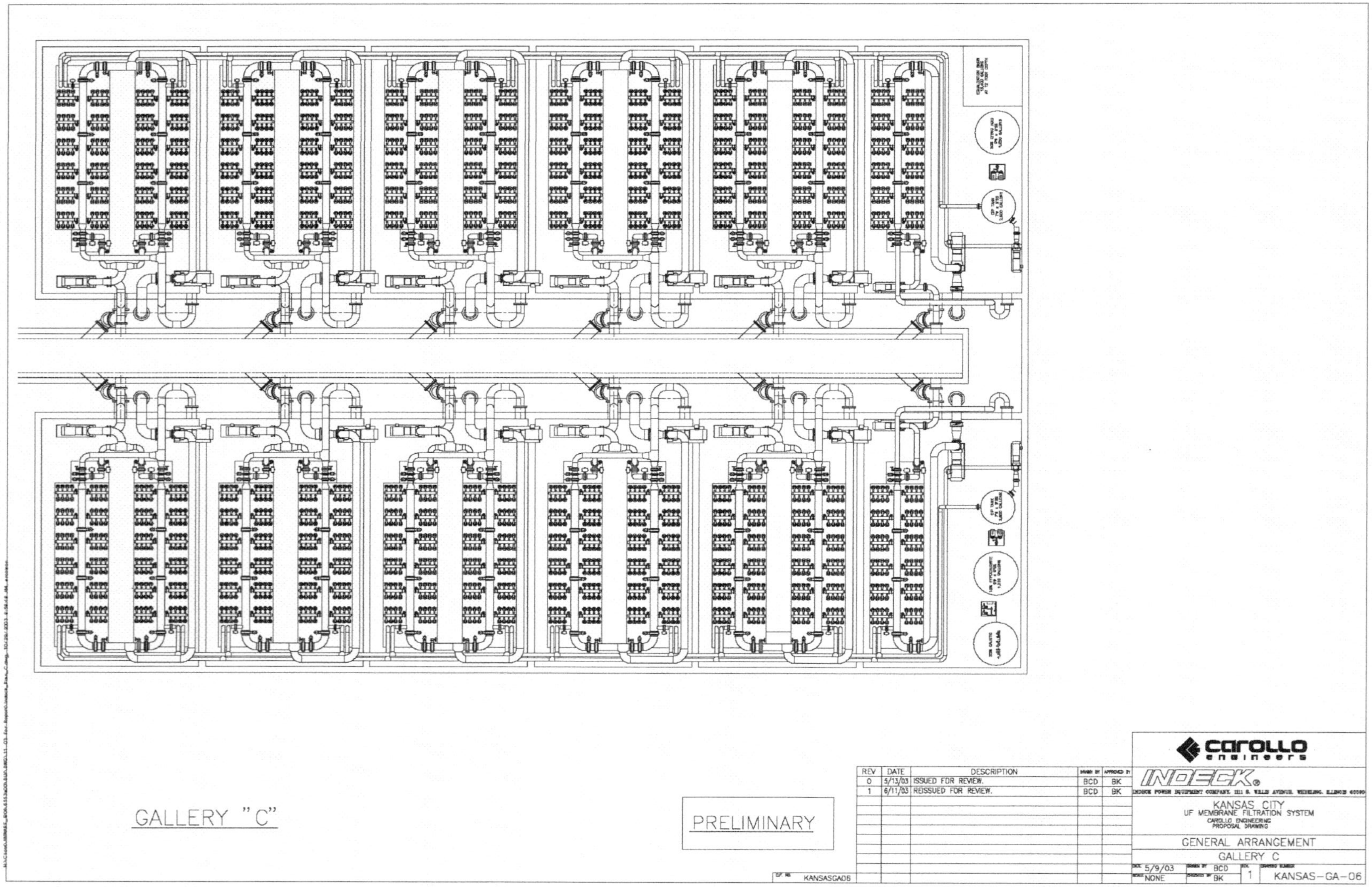

Source: INDECK Power Equipment Company and Carollo Engineers, P.C., 2003.

Figure 3.9 INDECK/Hydranautics system Scenario 2 layout

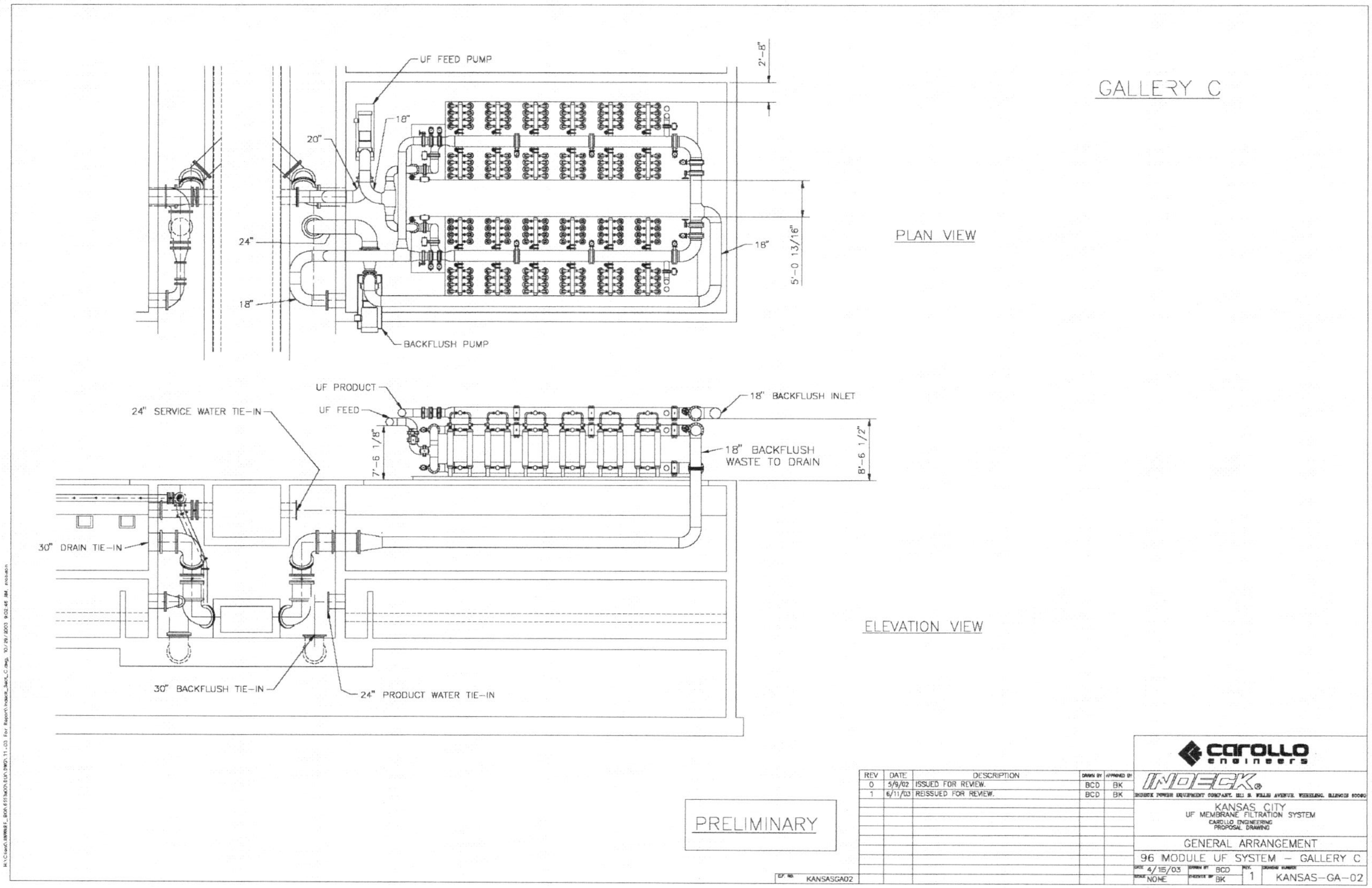

Source: INDECK Power Equipment Company and Carollo Engineers, P.C., 2003.

Figure 3.10 INDECK/Hydranautics system Scenario 2 cross sections

The required membrane racks fit into a total of 33 filter cells. One filter cell in each gallery is left available for ancillary equipment including chemical storage, recirculation pumps, dosing pumps, day tanks, and CIP flow equalization. Additional space may be available below the membrane racks if required. Dedicated backwash pumps are provided that operate inline with the existing high service pumps, which provide the necessary feed flow and pressure for the backwash. The racks were designed such that both halves of the rack can be operated independently of one another. Feed pumps are sized to provide flow to one half of the rack. Backwashing can therefore be performed as needed without the potential for extended backwash intervals. Based on the time required to backwash individual membrane racks and the number of racks, high service pumps could only be used for backwashing if multiple cells were backwashed simultaneously. In this case, the pump characteristics would allow this mode of operation.

Section views of the layout show the piping interconnections, including reuse of the existing washwater discharge line connections to the washwater sewer channel and the backwash supply line. The feed water connection is made at the existing feed channel penetration. The backwash supply connection is made at the existing filter supply cross. New piping is provided to connect these streams to the membrane rack.

Backwash discharge is routed away from the feed water header, underneath the membrane rack, and connected into the existing discharge piping. CIP solutions are introduced through the feed side of the membrane and are recirculated through a loop by a pump located in one of the forward filter cells. CIP waste is collected in an equalization/neutralization tank and discharged through the existing washwater sewer.

Westech/Polymem Layout for Scenario 2

Figure Nos. 3.11 and 3.12 show typical plan and section views of the Westech/Polymem layout for Scenario 2, respectively. As with the INDECK/Hydranautics system, the flux tested and demonstrated to meet the performance criteria for this project under the Scenario 2 conditions is nearly double that of the flux that met Scenario 1 conditions. Consequently, the number of membrane modules was reduced from 7,200 to 3,888 allowing the required membrane area to be installed in 24 of the 36 available filter cells. The number of modules per cell is decreased from 240 (Scenario 1) to 162 (Scenario 2) in this configuration. This leaves four filter cells per gallery for ancillary equipment including CIP recirculation pumps, compressors, chemical storage tanks, dosing pumps, and CIP waste equalization tanks.

As with the INDECK/Hydranautics layout for Scenario 2, the Westech/Polymem system includes a feed pump and a backwash pump. As with Scenario 1, it may be possible to utilize the high service pumps for backwashing, if a feasible alternative for rehabilitation of the backwash piping is identified and pressure surges can be mitigated. However, dedicated backwashing stations are assumed in the cost analysis (Chapter 4).

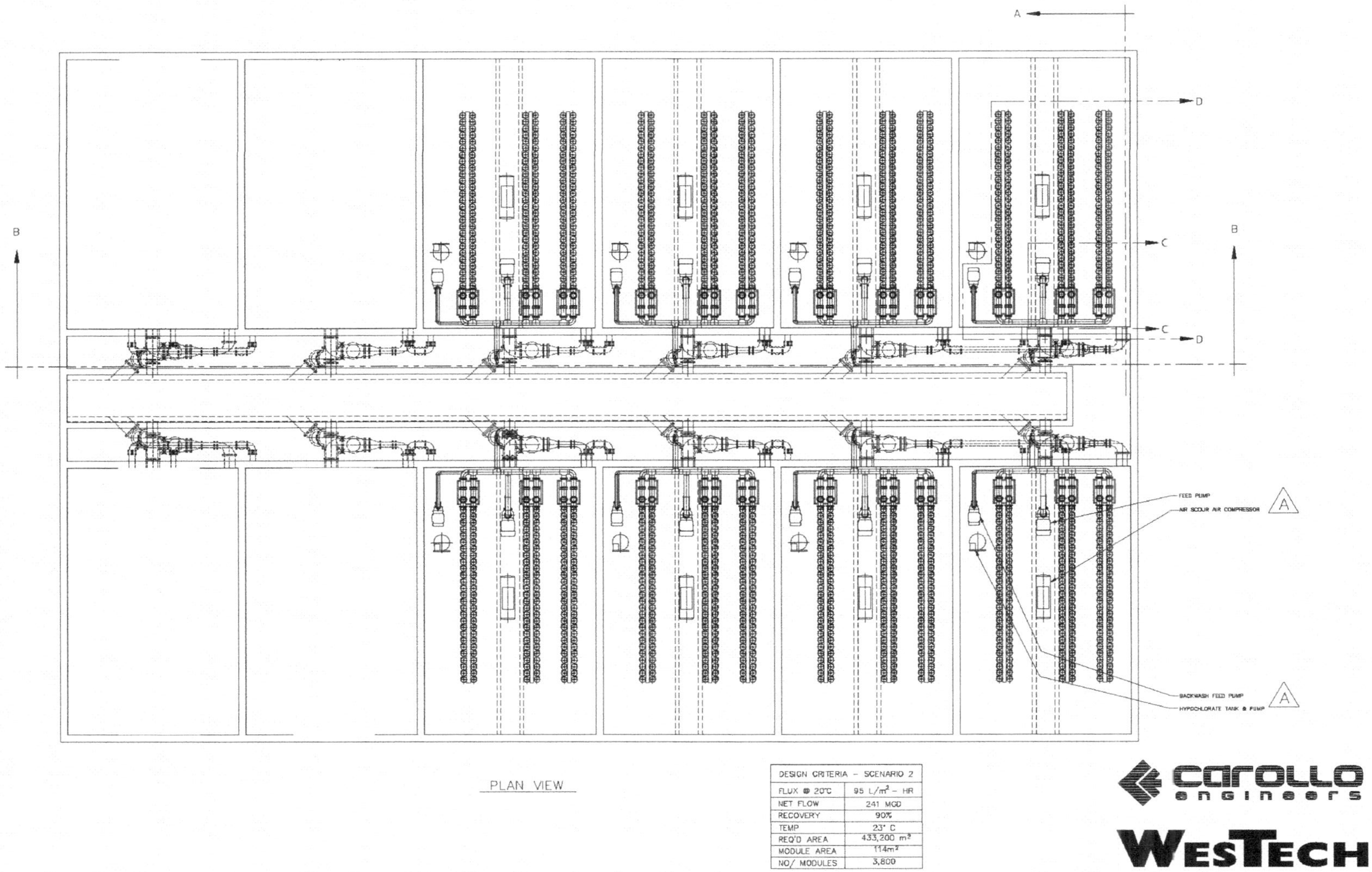

DESIGN CRITERIA – SCENARIO 2	
FLUX @ 20°C	95 L/m² – HR
NET FLOW	241 MGD
RECOVERY	90%
TEMP	23° C
REQ'D AREA	433,200 m²
MODULE AREA	114m²
NO/ MODULES	3,800

Source: Westech Engineering and Carollo Engineers, P.C., 2003.

Figure 3.11 Westech/Polymem system Scenario 2 layout

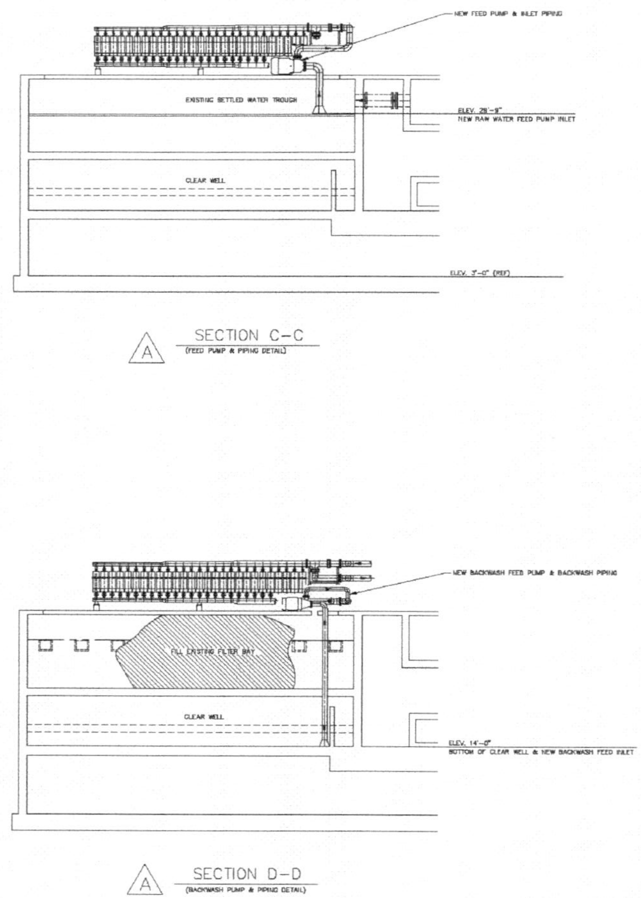

Source: Westech Engineering and Carollo Engineers, P.C., 2003.

Figure 3.12 Westech/Polymem system Scenario 2 cross sections

Zenon Environmental Layouts for Retrofit Scenario 2

Figures 3.13 and 3.14 are typical plan and section views, respectively, for the Zenon retrofit in Filter Galleries A and B. Due to the packing density of this immersed membrane configuration, the plant capacity is met using only one half of each filter cell to house membranes, the other half of the cell is solely dedicated to the permeate pump. Spare mounts within each process tank of the filter gallery provide space for additional membranes.

In this case, ten cells in each gallery are utilized for the membrane retrofit. Dedicated backwash and CIP solution tanks are located in the forward cells, as well as the accompanying pumps. The other forward filter cell is used for chemical storage and to house the required air compressors and blowers.

The system layout includes a new backwash distribution, air scour, and CIP recirculation header. Dedicated CIP and tank drain waste lines are also shown. Flow measurement and control is accomplished downstream of the permeate pumps within the filter cell. The existing filter rate of flow controller and backwash supply piping is abandoned; however, the existing clearwell penetration is utilized. Feed water is routed from the existing channel penetration through the lower gullet of the existing filter to the membrane tanks. CIP solutions are introduced into the system in a segregated portion of the membrane tank and enter the recirculation loop through the lower gullet. Tank drain and CIP solution waste is disposed of through this same connection to the washwater sewer.

Backwashing with the high service pumps may be easiest for this scenario. The ratio of production to backwash flux is low for this system (approximately 1.3), and the associated pressures are within the range of operational limits of the existing backwash supply lines in their current condition (< 20 psi). As with the other configurations presented, this approach would have to be evaluated by the supplier to ensure that pump and system characteristics will not damage the membranes.

CONSTRUCTION STAGING

The retrofit concepts described in this chapter require additions to and modifications of filter cells, piping, channels, and electrical infrastructure. In order not to interrupt production, construction would have to be staged.

The existing filter infrastructure is divided into three galleries, each with a capacity of 80 mgd. Existing infrastructure allows each filter cell to be isolated from a common gallery settled water channel and each cell has an isolation valve for backwash water. This would allow select cells to be removed from production or groups of cells during peak demand periods. An alternative to working within an active filter gallery would be to conduct the work in individual galleries from October through May. During this period maximum daily water demand can be met with two of the three galleries operating at less than full capacity. Work during the summer months would then be restricted to outside of the operating filter gallery.

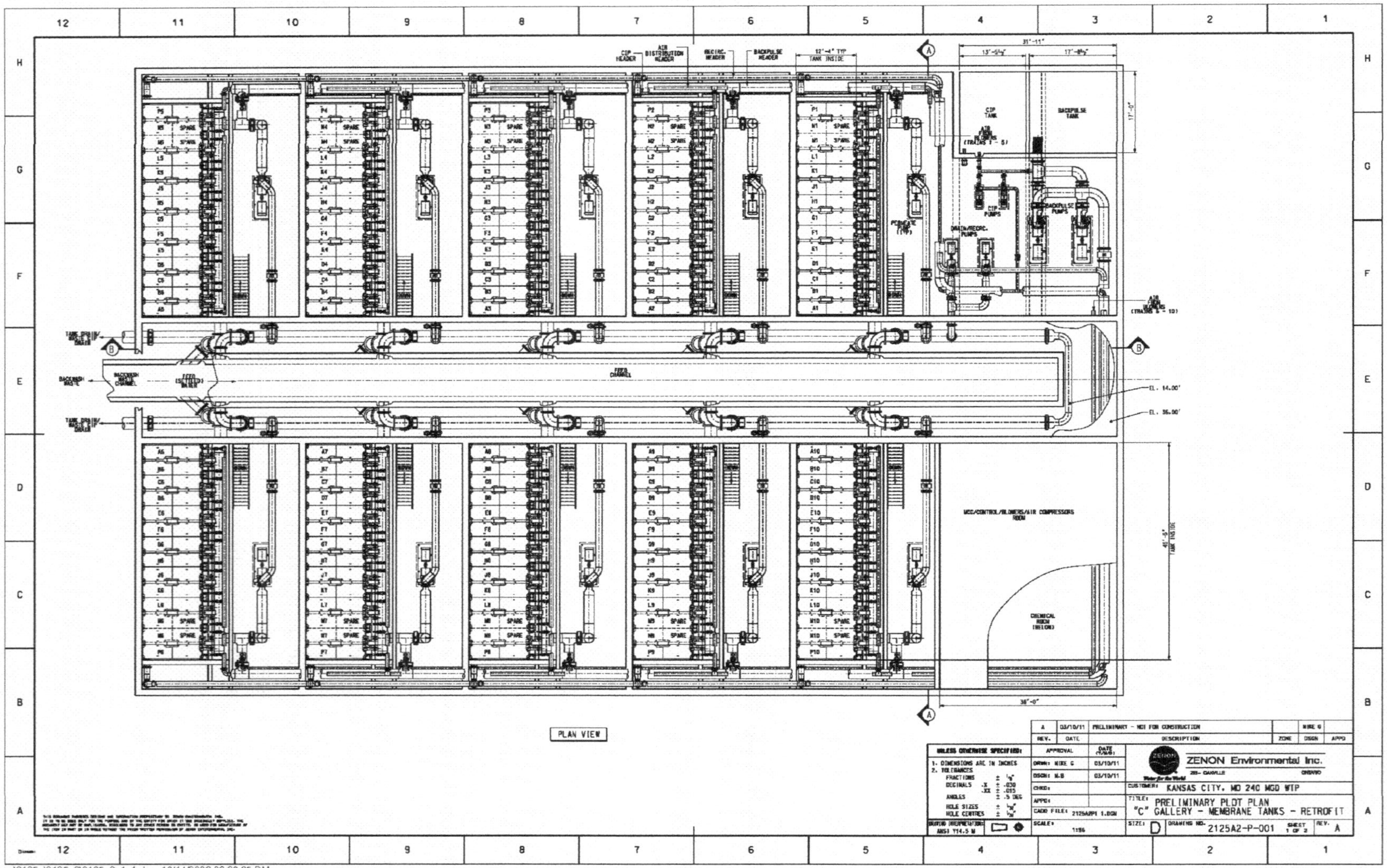

Source: Zenon Environmental, 2003b.

Figure 3.13 Zenon system Scenario 2 layout

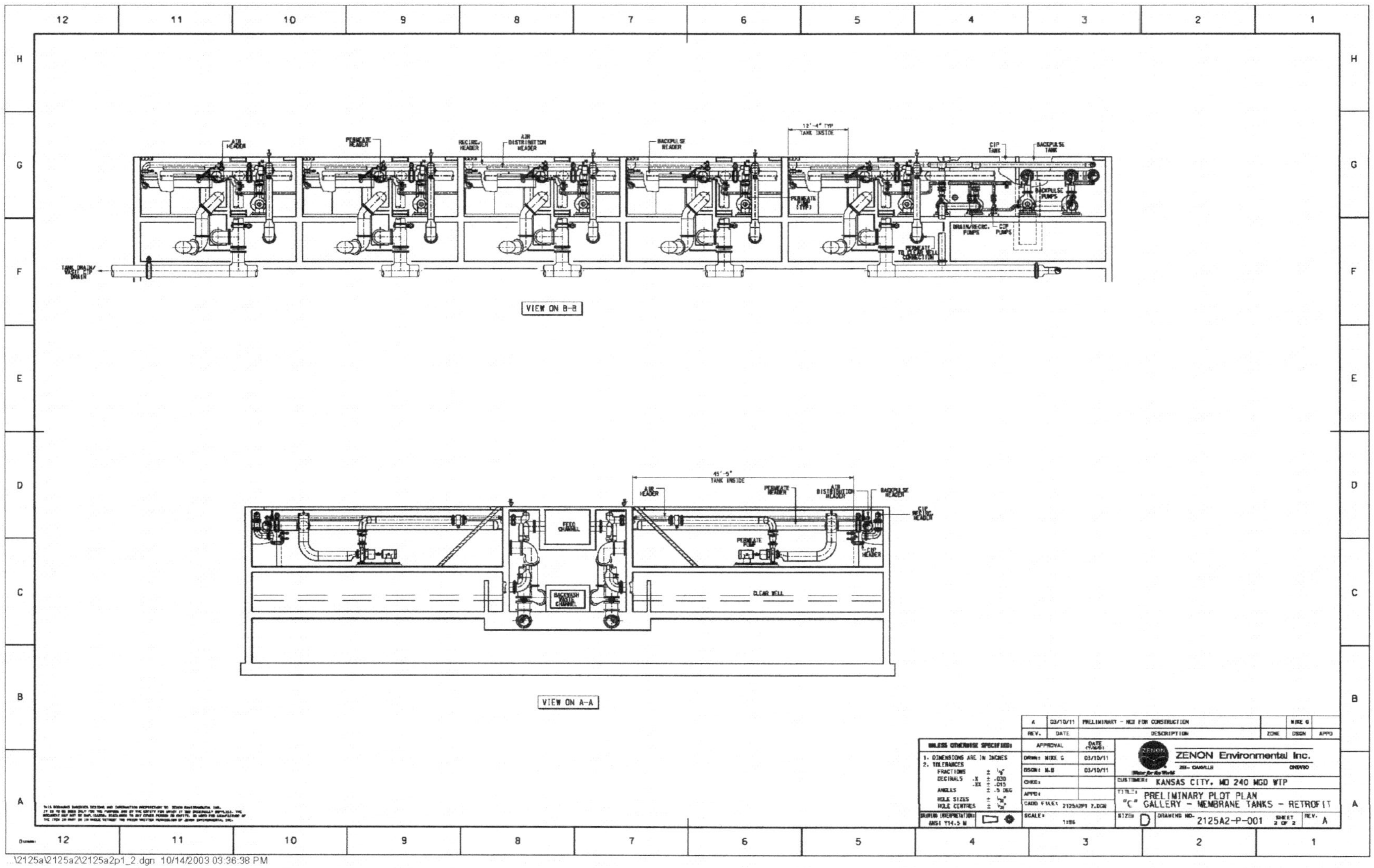

Source: Zenon Environmental, 2003b

•

Figure 3.14 Zenon system Scenario 2 cross sections

CHAPTER 4
RETROFIT COSTS

Estimated costs for the retrofits described in Chapter 3 at the KCWSD WTP are provided in this chapter. Critical capital cost components and lifecycle costs are provided. These include membrane racks and modules, structural/architectural, mechanical, electrical work, power, chemicals, and membrane replacement. The cost estimates were based on information from several sources including membrane supplier budgetary quotations, unit cost catalogs, and published and unpublished project costs for new and retrofitted plants. Life cycle costs are presented as a net present value (NPV) over a 20-year period. The assumed interest rate is 6 percent, and the corresponding present worth factor is 11.47.

Costs are given in ranges for the critical components identified for the retrofit based on scenario rather than supplier. This was done for several reasons, including market sensitivity of the cost data, conceptual nature of the project, and the lack of detailed plans and specifications. All costs are presented as 2003 USD.

SCENARIO 1 – CAPITAL COSTS

Scenario 1 cost estimates were based on the custom membrane system concept. Based on projections from the membrane supplier, 7,200 membrane modules were required to meet capacity requirements in this configuration. Membrane racks required 30 of the 36 available filter cells, and the forward two cells in each of the three galleries were used for storage of supporting equipment. The total estimated capital cost for this retrofit was $87.4 M ($0.364/gpd).

For this scenario, a fully integrated membrane system was assumed. The critical assumptions for this scenario included:

- No feed water pumping.
- Backwash supply provided by existing high service pumps.
- Backwash disposal through the existing filter gullet.
- All equalization, CIP, and ancillary equipment were located in the forward filter cells.
- Dedicated structures outside of the limits of the filter gallery for membrane racks and ancillary equipment were not required.

Membrane Racks and Modules

Costs for membrane modules and racks were based on estimates provided by the equipment supplier. These data, along with the rack designs presented previously, were used to develop budgetary estimates for the installed membrane racks and modules and its integral components, including instrumentation and control panels. The estimated installed cost for membrane racks and modules was $43.1 M ($0.179/gpd). Membrane racks and modules accounted for 49.4 percent of the estimated total capital cost.

Mechanical

Mechanical work included feed and backwash supply manifolds, CIP loops, air distribution, and associated valves that were not part of the membrane supplier scope. Feed water screening was also included in mechanical costs. The estimated cost for mechanical work described here was $9.6 M, 11.0 percent of the estimated total capital cost.

Structural

Major structural work for the retrofit included demolition of filter cell components, conversion of the forward cells to house ancillary equipment and chemical systems, and support structures for feed water screening equipment. The average estimated cost was $3.5 M, or 4.0 percent of the total estimated capital cost.

Electrical, Instrumentation, and Control

Major electrical work cost components for Scenario 1 included power to CIP pumps, compressors, electrical distribution, and instrumentation and control. The capacity of the existing electrical infrastructure would need to be evaluated to determine if the required loads could be accommodated. Instrumentation and control costs are a function of the level of integration with the existing control and monitoring infrastructure. In order to estimate costs for electrical work, a fixed percentage of the total project cost of 10 percent was used. Based on a review of new large membrane projects (> 15 mgd), electrical costs typically were found to be approximately 20 percent of the total project costs. Because existing electrical infrastructure, monitoring and control system already exist, and no significant additional pump capacity will be installed, a smaller percentage was selected for this cost evaluation.

Installed cost for electrical, instrumentation, and control systems was estimated to be $6.2 M, or 7.1 percent of the total estimated capital costs.

Contractor Overhead and Profit

It was assumed that contactor overhead and profit would be approximately 10.8 percent of the estimated construction cost.

General Conditions

It is assumed that the cost for compliance with general contract terms by the contractor will be approximately 7.1 percent of the construction subtotal.

Contingencies

A 15 percent contingency cost for the project was assumed, based on the subtotal estimated construction cost. This is equal to 10.8 percent of the total estimated construction cost.

SCENARIO 1 – OPERATION AND MAINTENANCE COSTS

Major lifecycle cost components for the system included membrane replacement, chemicals, and power. Membrane replacement costs were based on estimated module costs for the initial installation. (It should be noted that actual costs are based on negotiated arrangements between the owner and supplier and may vary significantly from initial purchase price.) Power costs were based on backwashwater pumping pressures and flow used in the pilot study. Chemical costs were based on chemical consumption estimates from the pilot study for a 30-day CIP interval. It was also assumed that the plant was to be operated at peak capacity (223 mgd at 20°C) year round.

Membrane Replacement Costs

Membrane replacement costs were based on costs for the initial facility installation, over a 20-year evaluation period. Replacement intervals used to calculate costs were based on 5- and 10-year replacement intervals. For the 5-year replacement interval, full inventory replacements were assumed at 5, 10, and 15 years. For the 10-year replacement interval, a single full inventory replacement was assumed at year 10. The NPV for membrane replacement was calculated using Equations 4.1 and 4.2:

$$NPV_{5-year} = \frac{C}{(1+i)^5} + \frac{C}{(1+i)^{10}} + \frac{C}{(1+i)^{15}} \quad (4.1)$$

$$NPV_{10-year} = \frac{C}{(1+i)^{10}} \quad (4.2)$$

where C = Number of membrane modules in system (No. modules) times single module replacement cost ($/module)

i = Interest rate (6%)

The estimated net present value for membrane replacement based on these assumptions for a 5-year replacement interval was $43.4 M. Assuming a replacement interval of 10 years, the membrane replacement cost was estimated to be $14.1 M.

Power Use and Costs

Power use for the membrane system in Scenario 1 was governed by backwash pumping and air compressor operating costs. The assumptions for this calculation are listed in Table 4.1. The estimated NPV for pumping power was $6.0 M.

Table 4.1
Assumptions for Scenario 1 energy cost estimate

Parameter	Assumed value	Note
Backwash pressure	35 psi	Max. pressure plus 3 psi for friction losses
Backwash flow rate	100 gpm/module	As demonstrated in pilot study
Pumping efficiency	65%	Motor efficiency x pump efficiency
Air flow	2 scfm/module	As demonstrated in pilot study
Compressor conversion factor	0.25	hp/scfm
Backwash duration	2 minutes	As demonstrated in pilot study
Backwash interval	50 minutes	As demonstrated in pilot study
Cost of energy	$0.08/kWh	

The net present value of pumping energy was based on Equations 4.3, 4.4, and 4.5:

$$kW = \frac{PQ}{1714} x\, 0.7457 \frac{kW}{\text{Hp}} \tag{4.3}$$

$$\text{Annual cost} = kW\ x\ t_{BW} x \frac{Cost}{kWh} \tag{4.4}$$

$$\text{NPV} = \textit{Annual}\ \text{Cost}\ x\, (1+\text{i})^{20} \tag{4.5}$$

The NPV of compressor energy was calculated using Equation 4.6.

$$AnnualCost = kW\ x\ t_{BW} x \frac{Cost}{kWh} x\, 365 \frac{days}{year} \tag{4.6}$$

where $Q_{air\text{-}STP}$ = No. modules x Air flow per module (SCFM)
kW = Compressor conversion factor (Hp/SCFM) x $Q_{air\text{-}STP}$
t_{BW}= Daily duration of backwash events (hr)

Chemical Consumption and Costs

Chemical consumption costs included chemicals used for backwashing, CIP events, as well as neutralization. Assumptions used in this calculation are provided in Table 4.2. It was assumed that one acid and one hypochlorite clean would be performed each month. The estimated NPV for cleaning chemicals based on the pilot data collected was $4.3 M.

Table 4.2
Assumptions for Scenario 1 chemical cost estimate

Parameter	Assumed value	Note
Backwash chemicals		
Recovery	90%	
Daily waste volume	27 MG	At peak flow
Hypochlorite dose	5 mg/L	
Cost of NaOCl	\$0.65/lb of Cl_2	Based on 100% active ingredient
CIP chemicals		
Cleaning frequency	30 days	
Cleaning system volume	100,000 gallons	
HCl concentration	500 mg/L	
HCl Cost	\$0.31/lb	Active ingredient
NaOCl concentration	200 mg/L	
NaOCl cost	\$0.65/lb	Active ingredient
NaOH (neutralization)	556 mg/L	1:1.11 mass ratio
NaOH cost	\$0.18/lb	Active ingredient
$NaHSO_3$ (neutralization)	280 mg/L	1:1.4 mass ratio
$NaHSO_3$ cost	\$0.26/lb	Active ingredient

The annual cost for cleaning chemicals was calculated using Equations 4.7 and 4.8.

For backwash hypochlorite:

$$\text{Annual Chemical Cost} = V_{BW} x C_{NaOcl} x 8.3454 \frac{lb/day/MG}{MG.mg/L} x \frac{Cost}{lb} x 365 \frac{day}{year} \quad (4.7)$$

where V_{BW} = daily volume of waste washwater (MG)
C_{NaOCl} = sodium hypochlorite concentration (mg/L)

For CIP chemicals:

$$\text{Annual Chemical Cost} = V_{\text{CIP Loop}} x C\ x 8.3454 \frac{lb/day/MG}{MG.mg/L} x \frac{Cost}{lb} x 12 \frac{CIP}{year} \quad (4.8)$$

where $V_{\text{CIP Loop}}$ = Volume (MG) of CIP Loop including tanks, membranes, and pipe volume
C = Concentration of cleaning or neutralization chemical

SCENARIO 2 – CAPITAL COSTS

Scenario 2 cost estimates are based on conceptual layouts and budgetary equipment estimates for a pumped feed water configuration. Due to the market sensitivity of the cost data, and at the request of one of the membrane suppliers, estimated costs for individual systems are not provided. The range of costs presented here are based on estimates from all three suppliers and included estimates for associated mechanical, structural, and electrical work. Total estimated capital costs ranged from $85.6 M to $105.1 M. On a capacity basis, the costs ranged from $0.355 to $0.438/gpd.

Membrane Racks and Modules

Costs for membrane modules and racks were based on estimates provided by the manufacturers. These data along with the rack designs presented previously, were used to develop budgetary estimates for the installed membrane racks and modules and its integral components. This included instrumentation and control panels. If not provided by the membrane supplier, costs for feed and backwash pumps were estimated and included in this cost element.

The estimated range of installed costs for membrane modules, racks, CIP equipment, and pumps for the retrofit is $33.2 M to $49.0 M ($0.138 to $0.204/gpd). Membrane rack and equipment costs account for 44.3 percent of the estimated total capital cost, on average.

Mechanical

Piping required to connect the membrane system to the existing clearwell, backwash supply, and backwash waste washwater handling systems were included in this section. Manifolds for these streams are also included in this cost. Costs were developed from piping and control concepts shown on the layout drawings. Mechanical costs were on average 7.4 percent of the estimated total capital cost ($7.1 M to $9.0 M).

Structural

Structural requirements for the retrofit included demolition of filter cell components, architectural considerations for filters following demolition, construction of membrane rack platforms (pressure-driven systems), and costs for conversion of the forward cells to dry pits for storage of ancillary equipment, chemical systems, and equalization tanks. Structural costs on average were 5.5 percent of the estimated total capital cost ($4.5 M to $5.5 M).

Electrical, Instrumentation, and Control

Major electrical work cost components for Scenario 2 included electrical distribution, motor control centers, instrumentation and control to accommodate feed, backwash, CIP pumps, and compressors.

The capacity of the existing electrical infrastructure would need to be evaluated to determine if the required loads could be accommodated. Instrumentation and control costs were a function of the level of integration with the existing control and monitoring infrastructure.

Although electrical, monitoring, and control infrastructure exist at the plant, it was assumed for Scenario 2 that significant upgrades would be required to accommodate feed and backwash pumps. In order to estimate costs for electrical work, a fixed percentage (20 percent) of the estimated subtotal construction cost was used ($12.7 M). This is consistent with new large membrane projects (> 15 mgd).

Other Costs

Engineering, contingency, and contractor overhead and profit costs consistent with Scenario 1 were assumed.

SCENARIO 2 – OPERATION AND MAINTENANCE COSTS

Major lifecycle cost components for the system include membrane replacement, chemicals, and power. Membrane replacement costs were based on estimated module costs for the initial installation. Power costs were based on feed and backwash water pumping pressures and flow used in the pilot study. Chemical costs were based on a 30-day CIP interval, and CIP and backwash chemical concentrations.

Membrane Replacement Costs

Since a breakout of module costs was provided by only two of the suppliers, the range of membrane replacement costs is based on two of the three membrane systems. A range of costs is provided for 5- and 10-year replacement intervals on a 20-year NPV basis. The range of membrane costs for 5-year replacement intervals is estimated to be $22.0 M to $31.4 M. The range of costs based on 10-year membrane replacement intervals is $6.4 M to $9.1 M.

Power Use and Costs

Power use for the membrane system in Scenario 2 is governed by feed water pumping, backwash pumping, CIP water heating, and air compressor operating costs. The range of values used in the calculation of energy use for each membrane supplier was based on pilot testing conditions. These values are listed in Table 4.3. The estimated NPV for pumping power is $7.0 M to $14.3 M.

Chemical Consumption and Costs

Chemical consumption costs include chemicals used for backwashing, CIP events, and neutralization. It was assumed that one acid and one hypochlorite cleaning would be performed each month. The estimated NPV for cleaning chemicals based on the pilot data collected ranges from $1.3 M to $4.3 M. The use of sodium hypochlorite in each backwash accounts for 69 percent of chemical costs for the two systems that utilize this fouling control strategy. CIP solutions and neutralization of cleaning waste accounts for the remainder of the chemical costs.

Table 4.3
Assumptions for Scenario 2 energy cost estimates

Parameter	Assumed value	Note
Feed pumping		
Feed pressure	7 to 12 psi	Average during pilot study plus 3 psi
Feed flow rate	6 to 48 gpm/module	Corresponding to 240 mgd at 20°C
Pumping efficiency	85%	Motor efficiency x pump efficiency
Hours per day in production		
Cost of energy	$0.08/kWh	
Backwash pumping		
Backwash pressure	13 to 35 psi	Max. pressure plus 3 psi for friction losses
Backwash flow rate	7 to 100 gpm/module	
Pumping efficiency	65%	Motor efficiency x pump efficiency
Backwash duration	1 to 2 minutes	
Backwash interval	30 to 50 minutes	
Cost of energy	$0.08/kWh	

Summary of Retrofit Costs

Table 4.4 provides a comparison between the estimated construction costs for Scenario 1 and Scenario 2 retrofits. While detailed designs and cost estimates were not developed for each layout, the differences in costs can be explained in general terms based on layout characteristics. It should be noted prior to this discussion that not all alternate cost saving designs were investigated, and cost estimates were based only on the layouts presented in previous sections.

Membrane module and equipment costs were approximately equal for the two scenarios. Inherent differences in system operation and module cost, and the lack of membrane feed and backwash pumps in the system evaluated in Scenario 1 made it cost competitive with the pumped systems evaluated in Scenario 2, despite the lower flux rate. Mechanical costs for Scenario 1 were generally higher due to the headers and manifold piping required to accommodate the number of subunits within each filter cell. Due to higher flux rates, Scenario 2 systems had fewer modules than Scenario 1 systems, resulting in fewer individual units and required valves and less header piping. While both retrofit approaches required structural work in the forward filter cells to accommodate ancillary equipment, Scenario 2 retrofits required additional structural work to support membrane racks. The pressurized systems required a supporting platform for installation above the filter cell, while the immersed system required tank separation and support for equipment hung on cell walls.

Table 4.4
Estimated capital construction costs

	Scenario 1*		Scenario 2†	
	Cost	% of total	Cost	% of total
Capital costs				
Membrane equipment	$43.1 M	49.4%	$42.0 M	44.3%
Mechanical	$9.6 M	11.0%	$7.8 M	7.4%
Structural	$3.5 M	4.0%	$5.2 M	5.5%
Electrical, instrumentation, and control	$6.2 M	7.1%	$12.7 M	13.4%
Subtotal Construction Cost	$62.4 M		$67.7 M	
General conditions (10%)	$6.2 M	7.1%	$6.8 M	7.2%
Contractor overhead and profit (15%)	$9.4 M	10.8%	$10.1 M	10.7%
Contingencies (15%)	$9.4 M	10.8%	$10.1 M	10.7%
Total estimated construction costs	$87.4 M		$94.8 M	

* Cost includes budgetary membrane equipment estimate from one supplier.
† Costs shown are averages of estimates developed for the 3 membrane systems for Scenario 2.

The line item with the largest difference between the two scenarios was electrical, instrumentation and control. Again, the assumptions used to generate these estimates were based on typical project costs, with consideration given to whether or not feed and backwash pumps were used. A detailed evaluation of electrical costs could result in different overall project costs for both scenarios.

O&M costs for the two scenarios (Tables 4.5 and 4.6) reflected the inherent differences in the layout concepts. Scenario 1 membrane replacement costs were significantly higher than those for Scenario 2. The design flux rates for Scenario 2 were 60 percent to 80 percent higher than for Scenario 1, resulting in fewer required modules and lower replacement costs. While the average energy cost for Scenario 1 was estimated to be 50 percent of the average Scenario 2 energy cost, the cost saving was insufficient to offset the difference in membrane replacement costs.

Table 4.5
Estimated operation and maintenance costs using 5-year membrane replacement intervals

	Scenario 1*		Scenario 2†	
O&M cost component	Cost	% of total	Cost	% of total
Membrane replacement	$43.4 M	80.9%	$26.5 M	63.6%
Energy	$6.0 M	11.2%	$12.0 M	28.9%
Chemicals	$4.3 M	7.9%	$3.1 M	7.5%
Estimated O&M costs	$53.7 M		$41.6 M	

* Cost estimated from Polymem UF120S3 module purchase costs, and performance data collected in the pilot study.
† Costs shown were developed from module purchase price information from two suppliers and individual system performance data collected during the pilot study.

Table 4.6
Estimated operation and maintenance costs using 10-year replacement intervals

	Scenario 1*		Scenario 2†	
O&M costs (20-year NPV)	Cost	% of total	Cost	% of total
Membrane replacement	$14.1 M	57.7%	$7.7 M	33.8%
Energy	$6.0 M	24.6%	$12.0 M	52.6%
Chemicals	$4.3 M	17.6%	$3.1 M	13.6%
Estimated O&M costs	$24.4 M		$22.8 M	

* Cost estimated from Polymem UF120S3 module purchase costs, and performance data collected in the pilot study.

† Costs shown were developed from module purchase price information from two suppliers and individual system performance data collected during the pilot study.

The difference in membrane replacement cost became less significant as the replacement interval increased. Because membranes were the largest component of the operation and maintenance costs, decreasing the number of replacement events by two (5-year versus 10-year intervals) amplified the energy cost component. In Scenario 2, membrane replacement was 63.6 percent of the O&M cost at 5-year replacement intervals and decreased to 33.8 percent on a 10-year replacement basis. Electrical costs increased from 28.9 percent to 52.6 percent of the total O&M costs for the same change in membrane replacement interval.

Cost amortization provided a means to compare costs between scenarios. Costs were amortized to water production by multiplying annual operating cost by a 20-year lifetime and adding that capital costs, and dividing by total water addition over 20 years, as shown in Table 4.7. Amortization assumes operation at peak capacity.

Table 4.7
Amortized project costs

	5-year membrane replacement interval	10-year membrane replacement interval
Scenario 1	$0.103/1000 gallons	$0.074/1000 gallons
Scenario 2	$0.096/1000 gallons	$0.077/1000 gallons

CHAPTER 5
PERSONNEL, START-UP, AND O&M CONSIDERATIONS

CERTIFICATION AND PERSONNEL REQUIREMENTS

Most states, including Missouri, do not require special certification for membrane-based WTP operation. However, since each state has its own water treatment plant operator certification program, water utilities will have to check with their local regulatory agency to determine if special membrane operation certification is required. As the integration of membranes into water treatment plants becomes more common, the implementation of a specialized membrane certification requirement may become standard.

In general, large WTPs will not require additional personnel to operate a membrane filtration system. In addition, the time required for regular operation of a membrane filtration process is comparable to that required for the operation of a granular media process. However, initial operator training is required to learn the operation, control, and maintenance issues associated with membrane filtration.

OPERATOR TRAINING

Typically, membrane suppliers conduct the majority of operator training sessions on-site once a membrane system has been installed. O&M manuals, provided by the membrane supplier, are used in conjunction with these sessions. Operators can also attend membrane seminars or training courses for which they could receive continuing education units. Occasionally, a few membrane treatment plants offer training courses that give operators hands-on experience with the daily operation of a membrane filtration system. Training programs specific to the operation of a membrane filtration process may include some or all of the following:

- Introduction to membrane systems and associated terminology. Operators will learn about flux, transmembrane pressure, specific flux, normalized flux, membrane permeability, CIPs, and other related concepts.
- System start-up and shut-down procedures.
- Long-term storage procedures.
- Backwash protocol.
- CIP protocol. This includes learning how to handle new chemicals and measure the concentrations of stock cleaning solutions.
- Integrity testing. Direct integrity testing can be programmed to occur automatically, or operators can initiate an integrity test manually. Direct integrity tests must be performed at least once per day if used to comply with the LT2ESWTR for *Cryptosporidium* removal credit, although some states may require integrity tests to be done more frequently.
- Fiber repair.
- Module replacement.
- System monitoring and control. This includes the use of PLCs and SCADA systems and may also involve learning how to calibrate high-resolution particle counters and turbidimeters. Most membrane control systems can be operated with various levels of control. Operators will be trained how to manage the system during fully manual,

semi-automatic, and fully automatic modes. Semi-automatic mode allows an operator to initiate an action, such as an integrity test or CIP, even though the system is normally operating automatically. Complete training in all modes of operation significantly increases the level of operator confidence and can improve trouble-shooting capabilities.

- Trouble shooting.

Proposal Draft version of the United States EPA office of Water's Membrane Filtration Guidance Manual (June 2003) does not include any specific requirements for operator training; however a sample training schedule is provided in the document. The document suggests 24 to 86 hours of supplier-provided training during start-up and commissioning, as well as 1 to 3 day open discussion periods following plant commissioning and three months thereafter.

Once the initial training is complete, a practice period is often implemented for plant operators, which can be supervised by the membrane supplier and/or associated engineering firm. This practice period can be used to expose operators to a variety of problematic scenarios, which they are likely to face, and gives them practice trouble shooting under low-risk conditions.

START-UP AND OTHER CONSIDERATIONS

Membrane modules are often shipped in a glycerin solution, which must be flushed and disposed. Because the solution has a high BOD, wastewater treatment plants must be notified before it is discharged to the sewer. If the loadings are too high for the wastewater plant to handle, the solution may need to be diluted or discharged over the course of several days. If air gets trapped in the membrane pores during the flushing period, the membranes need to be rewetted prior to system start-up. This is done by recirculating a dilute caustic solution through the membrane modules for several hours.

New record keeping protocols may be necessary once a membrane filtration system is in operation. Additional water quality parameters, flows, and/or pressures may need to be recorded, although typically much of this is automatic. In addition to the direct integrity tests mentioned previously, water plants also are required to perform indirect integrity testing at minimum 15-minute intervals to comply with LT2ESWTR. Effluent particle counts are often used to indirectly monitor membrane integrity (USEPA 2003).

One other consideration for the integration of membranes in water treatment is the importance of proactive operation. Upstream processes and water quality must be monitored regularly to avoid fouling the membrane modules. Shifts in pH, and concentrations of coagulants, polymers, or other chemicals can rapidly deteriorate performance and the integrity of membrane systems, leading to the costly replacement of modules.

CHAPTER 6
CONCLUSIONS AND RECOMMENDATIONS

Increasing regulatory focus on the microbial water quality of drinking water has driven many utilities to consider advanced treatment technologies such as membranes. Membrane filtration is now being applied broadly in drinking water treatment to meet compliance challenges in new facilities. However, construction of a new facility is not feasible in some locations due to land availability, cost implications, or logistical concerns. These limitations become particularly visible at large treatment plants where constructing dedicated membrane filtration facilities and related infrastructure could cost hundreds of millions of dollars. One alternative to building new membrane facilities is to retrofit WTPs with membrane filtration, utilizing existing infrastructure, piping, and ancillary equipment. However, retrofits of granular media filters on the scale of the KCWSD WTP are unprecedented and implementation issues are undocumented.

While there were examples of membrane retrofits with varying degrees of integration at the time this document was written, they were smaller scale than that described herein and utilized less of the existing infrastructure. The systems tested and evaluated for retrofit in this project were selected to represent the range of options available for retrofits and included vacuum-driven immersed, standard pressure driven, and custom designed pressure driven membranes.

PILOT TESTING

Pilot testing in this study was used to provide design criteria used to develop retrofit layouts. The pressure driven systems were evaluated at low flux conditions that would minimize the required feed pressure to permit operation within the existing filtration hydraulic grade line of the plant. This required maximum TMPs to be less than 10 ft (4.3 psi). Both systems were able to meet minimum cleaning interval and recovery goals. Flux rates meeting this criteria ranged from 30 gfd at 20°C to 41 gfd at 20°C. High membrane permeability and/or low fouling rates are necessary to operate under gravity-fed conditions.

Testing data showed that under a pumped-fed scenario, flux rates for the pressure driven systems were up to 86 percent higher than the gravity-fed scenario. Under these conditions, none of these systems failed to meet the CIP interval or recovery goals set for the project. Maximum TMPs for the pressure-driven systems ranged form 22 to 29 psi. Maximum flux rates tested for the two membranes were 56 and 85 gfd at 20°C. At these flux rates both systems met cleaning and recovery goals, and higher flux rates could not be tested only due to pilot equipment limitation and project duration. The maximum terminal TMP for the vacuum-driven immersed system (10 psi) was used as evaluation criterion throughout the study. The maximum flux that consistently met the cleaning interval goal over the range of feed water conditions tested was 20 gfd at 20°C. At flux rates exceeding 20 gfd at 20°C and median feed turbidity greater than 9.9 NTU, the vacuum system did not meet the cleaning interval goals set for the project. Though lower recoveries may have decreased fouling, the system was operated at a constant recovery of 95 percent, as recommended by the manufacturer.

The potential impacts of carry-over of both PAC and filter-aid polymer on membrane performance were evaluated during the study. PAC loading onto the filters is observed regularly at the existing plant. In order to evaluate its impact on membrane performance, a complete run was performed with a PAC dose of 5 mg/L. The operating conditions selected for the PAC run were also tested without PAC. The total increase in TMP over the course of a run was less with the addition of PAC than without. There were no trend in feed or permeate water associated with these runs. It is likely that the PAC was removing some of the filter aid polymer present in the feed water, thus decreasing the fouling rate. It is also possible that the nature of the solids layer that developed on the membrane surface was altered by the PAC to be more readily removed by production backwashes.

Filter-aid polymer was added at 1.2 mg/L (10 times the normal operating level) to evaluate potential impact on membrane fouling. Two of the three systems were tested. The length of the event was set at 1 to 3 days until model system upset. Impacts on fouling and cleaning varied. While fouling was readily apparent for the INDECK/Hydranautics membrane, a hypochlorite clean followed by an acid clean completely restored membrane permeability. For the Westech/Polymem membrane, no significant increase in transmembrane pressure was observed following polymer addition. However, only partial membrane permeability was restored during the subsequent cleaning cycles.

RETROFIT LAYOUTS

Custom layouts were developed to maximize equivalent SLR around each of the membrane options for two distinct configurations, defined by design criteria developed for the study. The flux rates and recoveries used for these evaluations were selected based on the performance during the pilot study.

Scenario 1 was a membrane system which operates within the existing filtration hydraulic grade line during production, utilized the existing infrastructure for backwashing, and utilized the existing waste washwater gullet for backwash disposal. This approach required that the membrane racks and modules fit within the existing filter cells, and that sufficient space was reserved within the filter galleries for ancillary equipment and tanks. The key to successfully meeting the requirements of this retrofit approach was to exceed the net hydraulic loading rates of the existing filter cells.

Only the Westech/Polymem system met these criteria. The INDECK/Hydranautics system required the construction of a dedicated structure to house the required membrane racks and modules. Critical differences between Westech/Polymem and INDECK/Hydranautics membranes drove feasibility. These differences included; the number of feed and permeate ports, fiber packing density, and active surface area within each module. This reduced the space required for piping and valves and the number of modules and connections required.

Scenario 2 was a more typical approach to membrane retrofits. In terms of its equipment requirements, it utilized feed water pumps and dedicated membrane backwash stations. Feed pumps allowed higher TMPs and flux rates, resulting in fewer membrane modules, racks, valves, and less interconnecting piping. All three systems evaluated were compact enough to fit within the existing filter gallery under the conditions of Scenario 2. The number of filter cells used to house membrane racks varied from 24 to 33. The most compact layouts (highest equivalent SLRs) were designed around high fiber packing density modules, not around the modules with the highest flux. While this result is application-specific (flux rates are highly dependent on site-specific water quality), fiber-packing density should be given consideration when selecting the systems to be tested in retrofit projects of this type.

RETROFIT COSTS

Capital and life cycle cost estimates for both scenarios were developed. On an average capital cost basis, Scenario 1 was approximately 9 percent less than Scenario 2. At the lower end of the capital cost estimates for the systems evaluated in Scenario 2, capital costs were not significantly different from Scenario 1. Membrane equipment cost for both scenarios was approximately $0.17/gpd. While the number of modules and racks was larger for Scenario 1, the cost of the pumps required for feed and backwash in Scenario 2 was largely offset membrane cost savings.

Estimated mechanical costs for Scenario 1 were higher than for Scenario 2, due the costs of interconnecting piping and associated valves and fittings. Structural costs for Scenario 1 were lower, as modifications of filter boxes was minimized. Electrical costs were found to be the most significant cost difference between the two design scenarios. For Scenario 1, these costs are less since motor control centers, electrical distribution and infrastructure upgrades are minimal.

Membrane replacement drove life cycle O&M costs. At 5-year replacement intervals, Scenario 1 average membrane replacement costs were 64 percent higher than those for Scenario 2. This was due to the low flux rates required for a Scenario 1 retrofit. At longer replacement intervals, membrane replacement was a smaller fraction of life cycle O&M costs, and the energy savings for Scenario 1 became a more important component.

RECOMMENDATIONS

Because the project success of the retrofit will rely on the proper operation and modification of existing infrastructure, an audit of critical equipment should be performed. This includes compilation of existing studies at the plant as well as an assessment of the following (at a minimum) to develop additional information required for design and construction.

- Surge analysis (backwash line)
- Structural integrity of filter boxes and sedimentation basins
- Electrical infrastructure
- Control systems integration
- Longer term hydraulic performance study to evaluate cleaning effectiveness and impact of warmer water temperatures on membrane fouling under constant normalized flux operation.

APPENDIX A – PILOT TESTING DATA

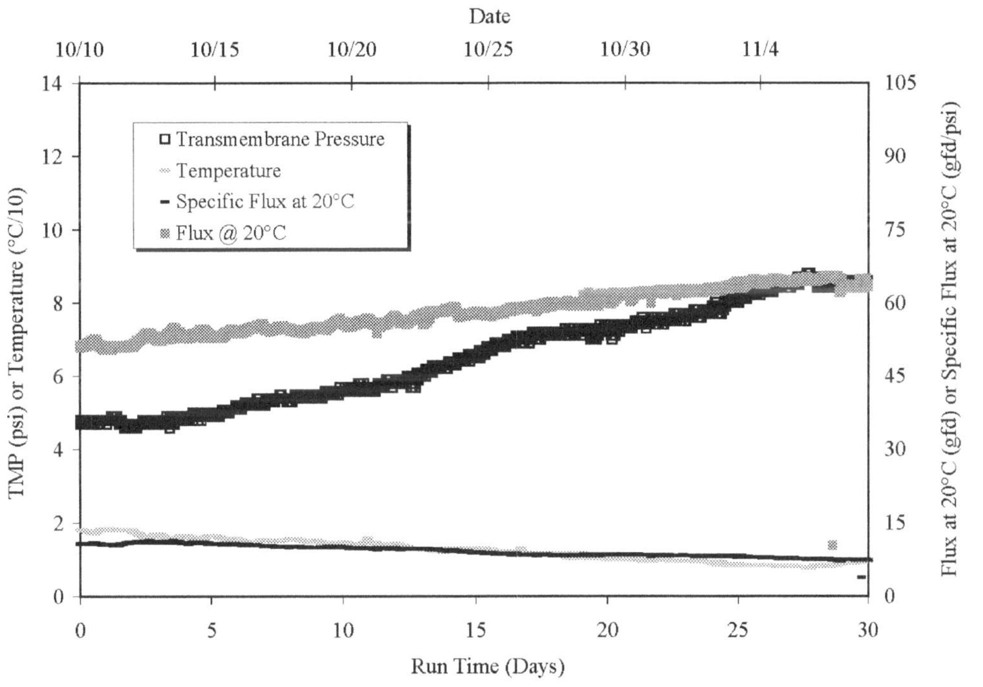

Figure A.1 Run 2 INDECK/Hydranautics hydraulic performance (operating conditions: average flux @ 20° C = 58.0 gfd, backwash interval = 40 minutes, recovery = 93%)

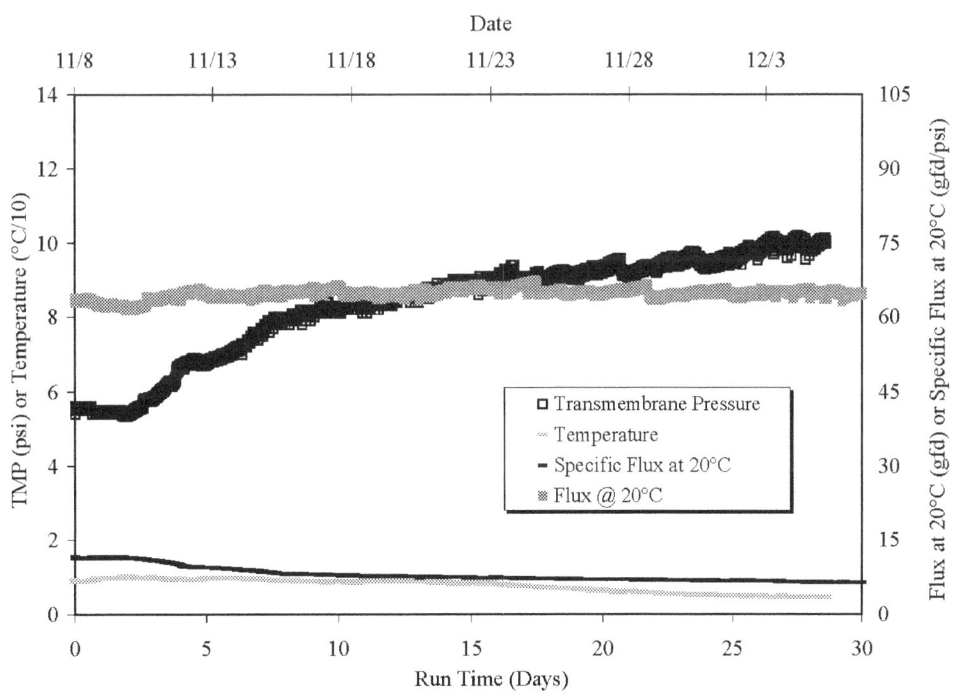

Figure A.2 Run 3 INDECK/Hydranautics hydraulic performance (operating conditions: flux @ 20°C = 65.0 gfd, backwash interval = 42 minutes, recovery = 93%)

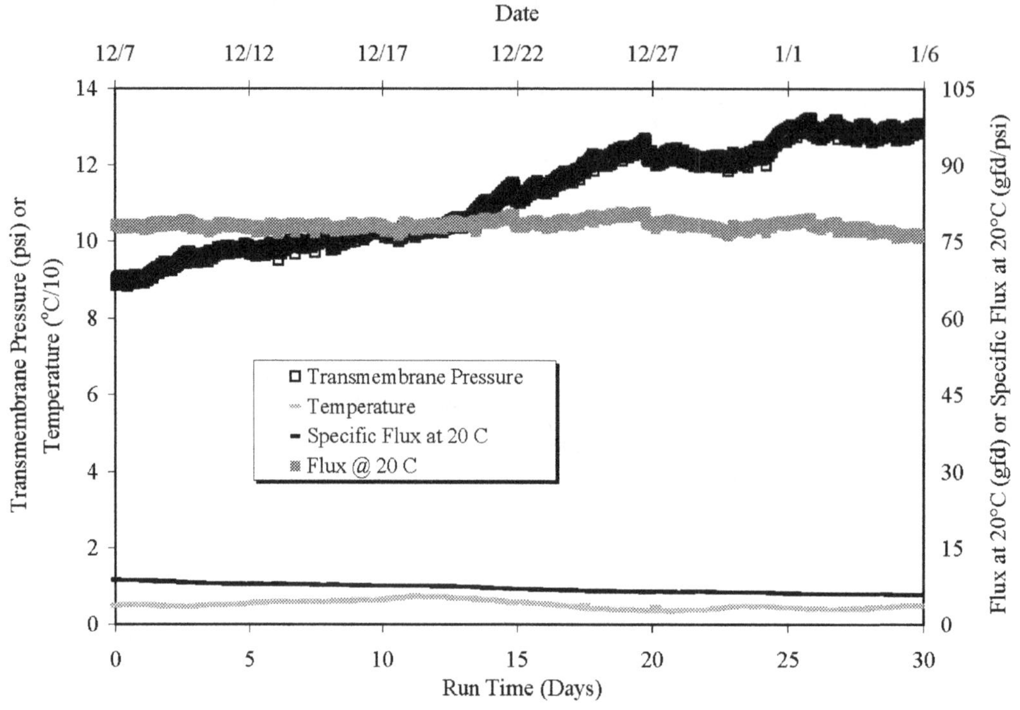

Figure A.3 Run 4 INDECK/Hydranautics hydraulic performance (operating conditions: flux @ 20°C = 78.0 gfd, backwash interval = 34 minutes, recovery = 93%)

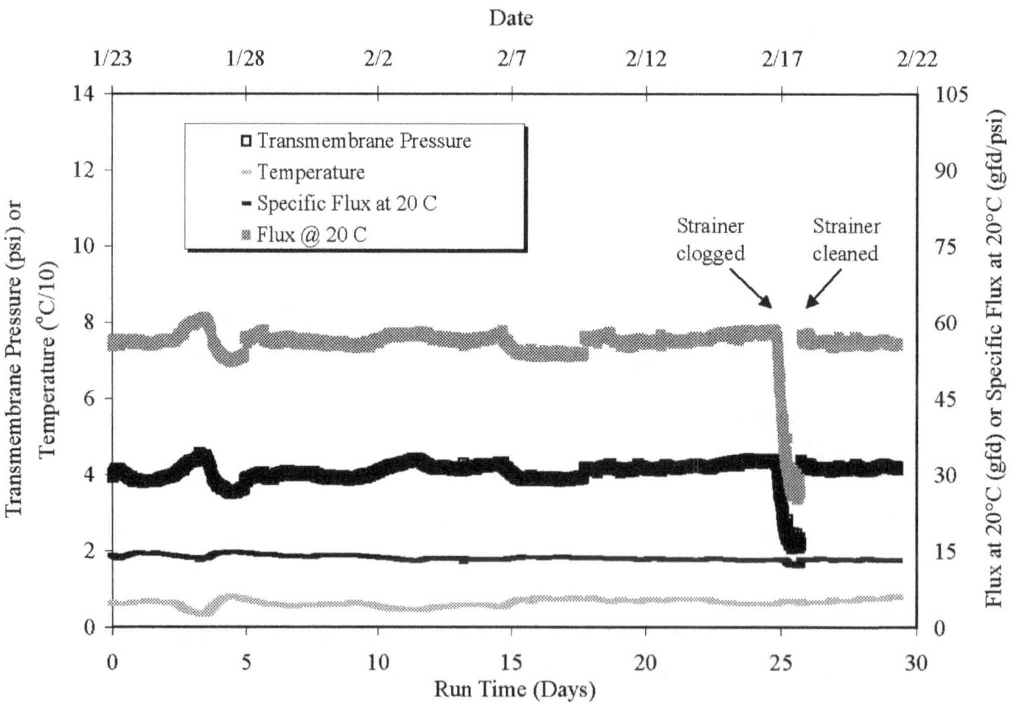

Figure A.4 Run 5 INDECK/Hydranautics hydraulic performance (operating conditions: flux @ 20°C = 58.0 gfd, backwash interval = 34 minutes, recovery = 90%)

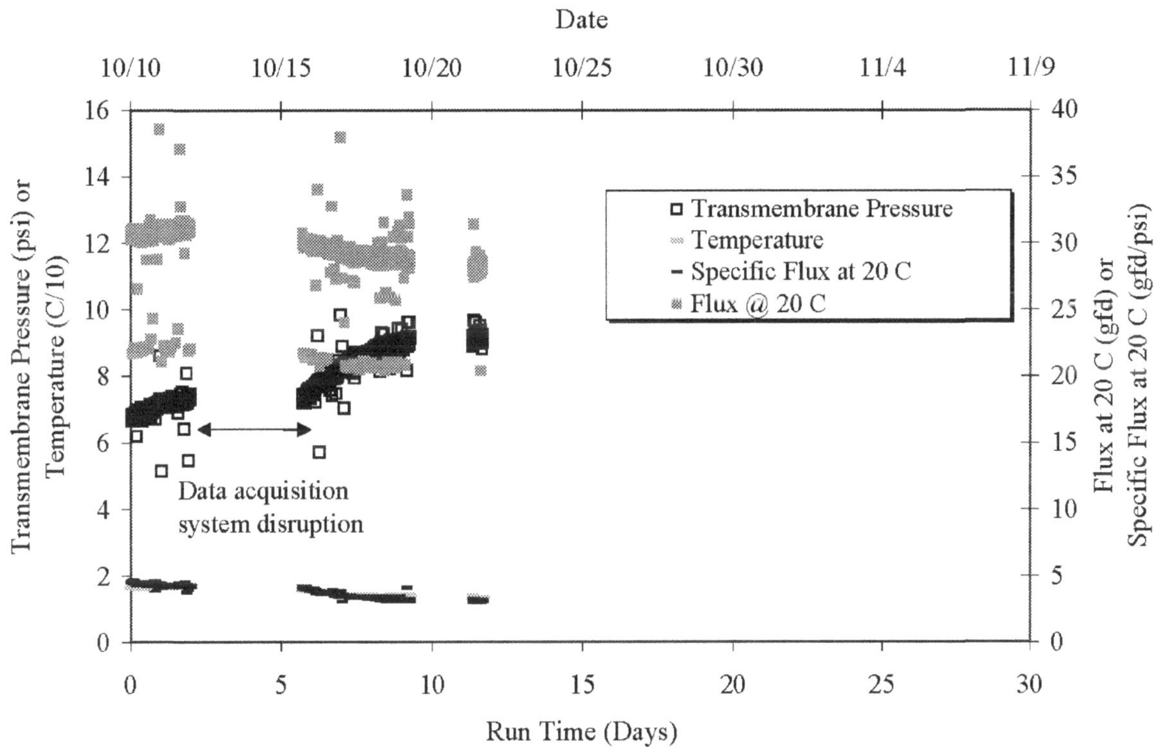

Figure A.5 Run 2a Zenon hydraulic performance (operating conditions: average flux @ 20°C = 36.9 gfd, backwash interval = 24 minutes, recovery = 95%)

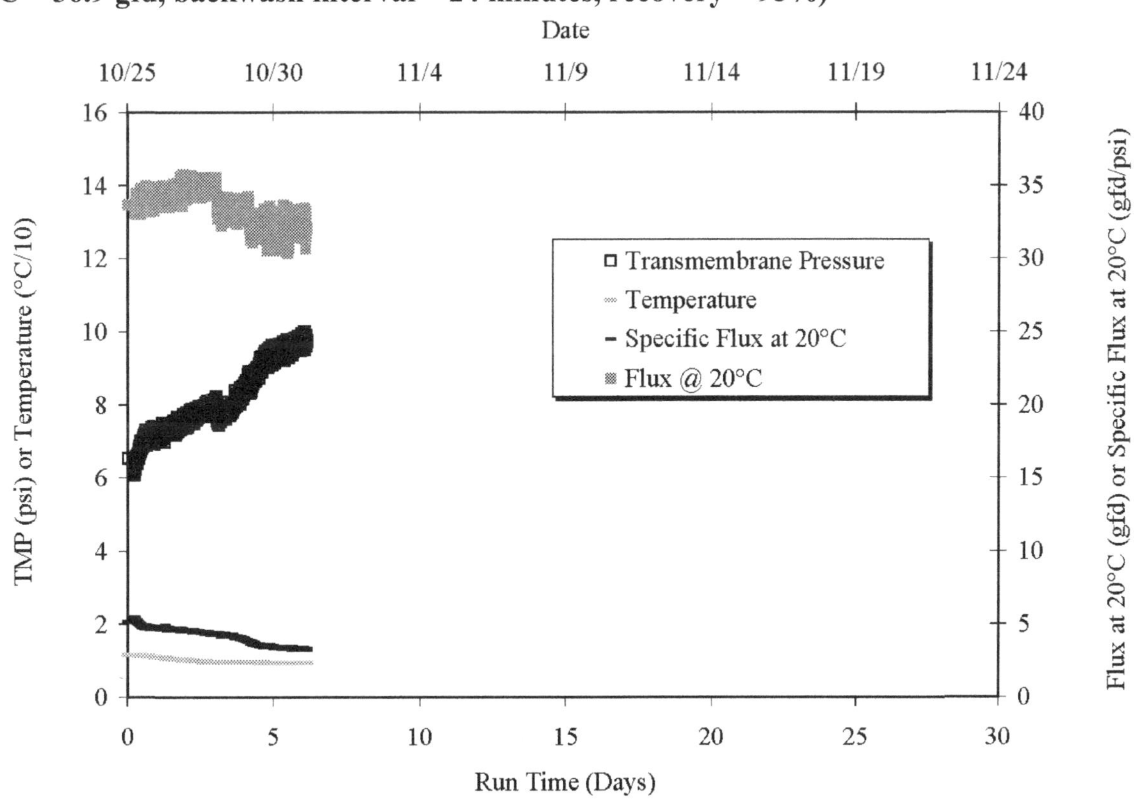

Figure A.6 Run 2b Zenon hydraulic performance (operating conditions: average flux @ 20°C = 33.2 gfd, backwash interval = 33 minutes, recovery = 95%)

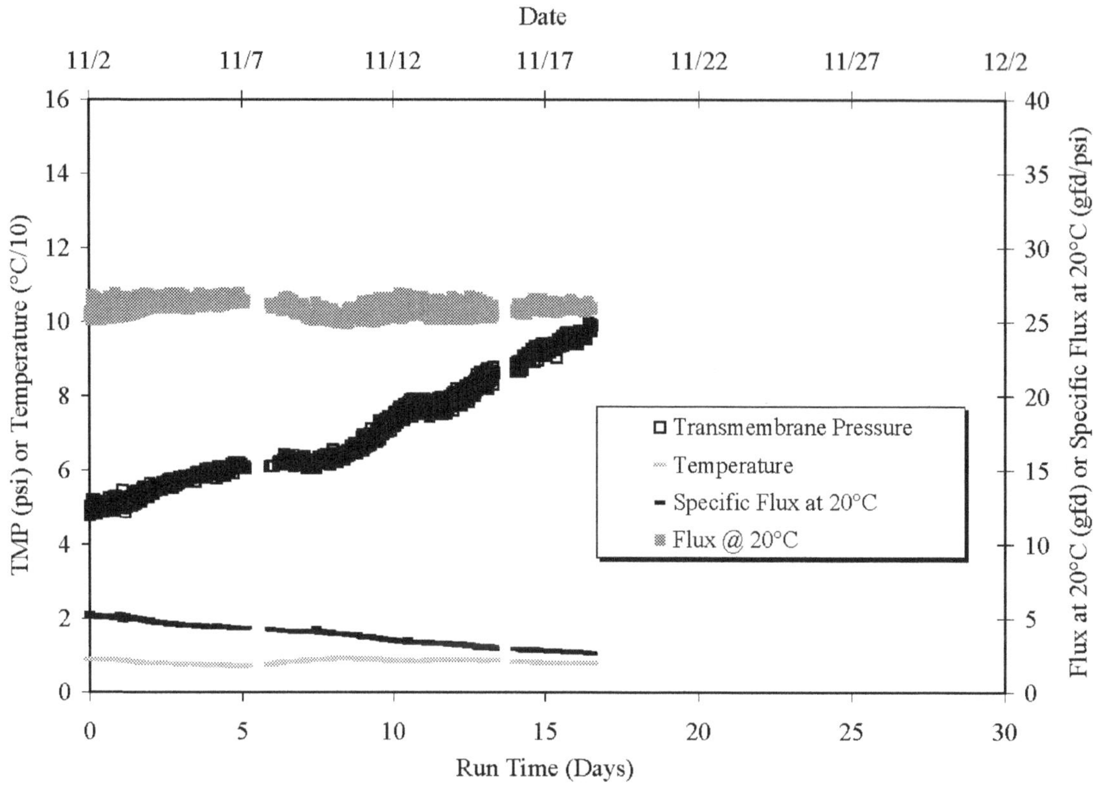

Figure A.7 Run 2c Zenon hydraulic performance (operating conditions: average flux @ 20°C = 26.4 gfd, backwash interval = 40 minutes, recovery = 95%)

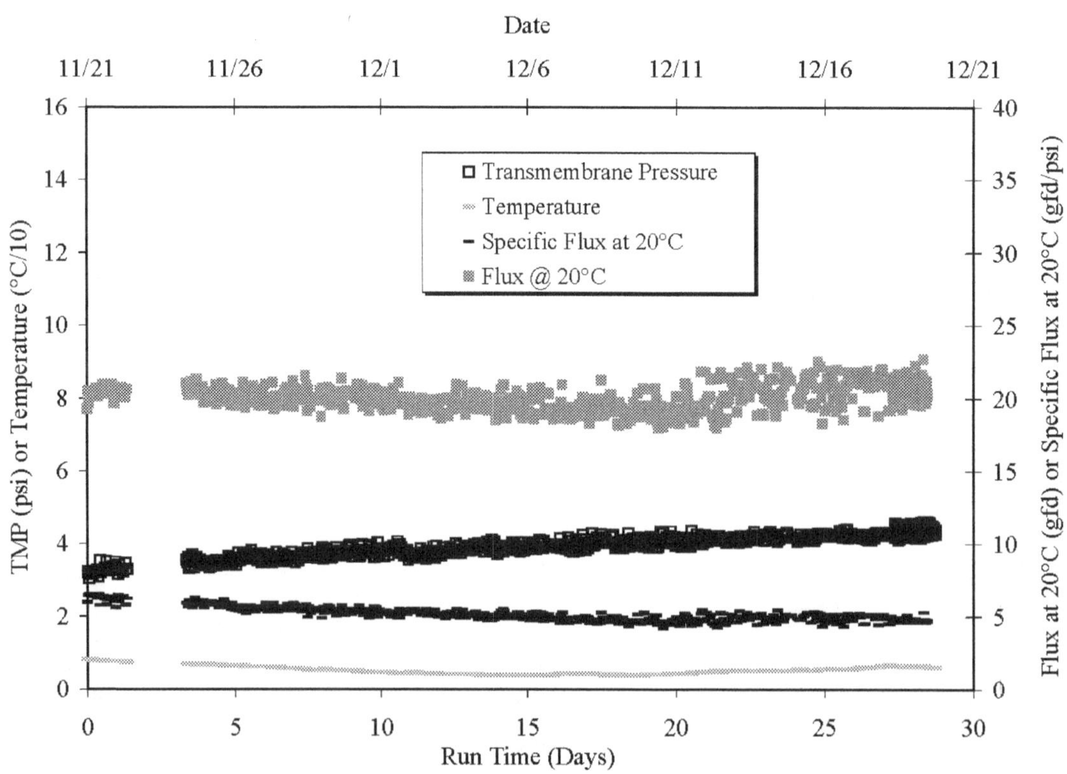

Figure A.8 Run 3 Zenon hydraulic performance (operating conditions: flux @ 20°C = 20.0 gfd, backwash interval = 54 minutes, recovery = 95%)

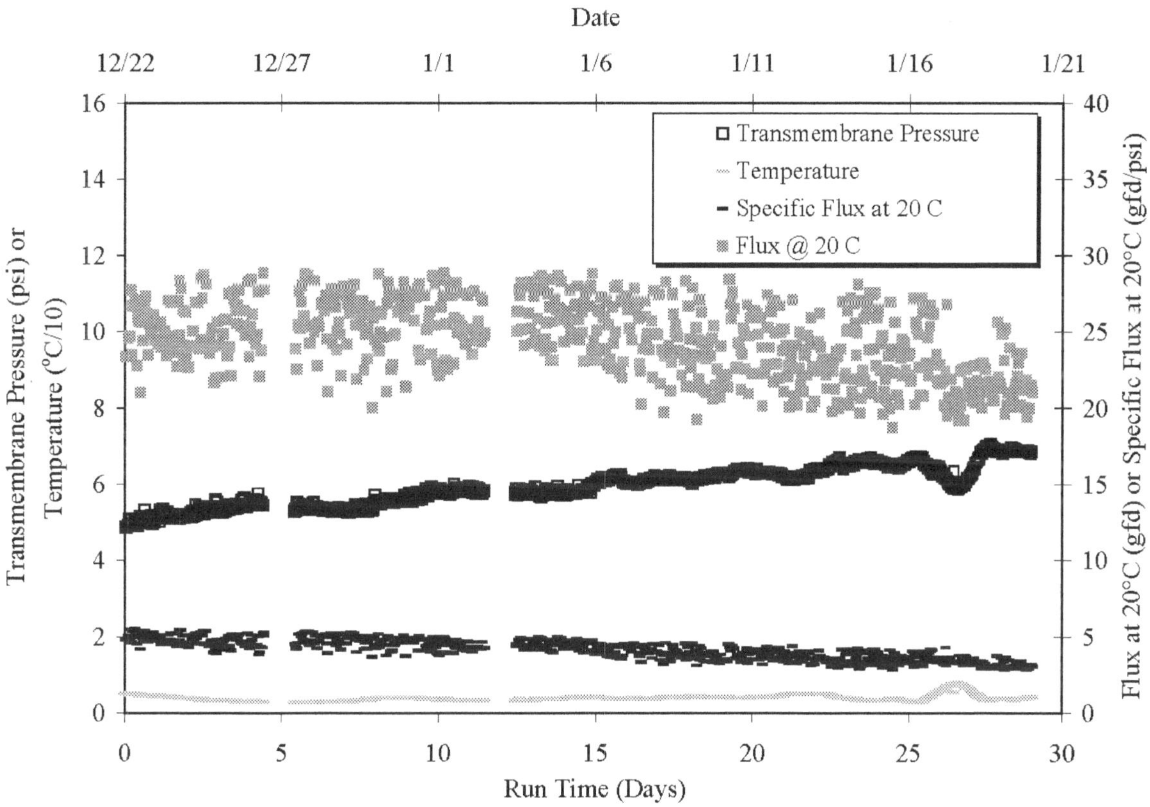

Figure A.9 Run 4 Zenon hydraulic performance (operating conditions: flux @ 20°C = 25.0 gfd, backwash interval = 45 minutes, recovery = 95%)

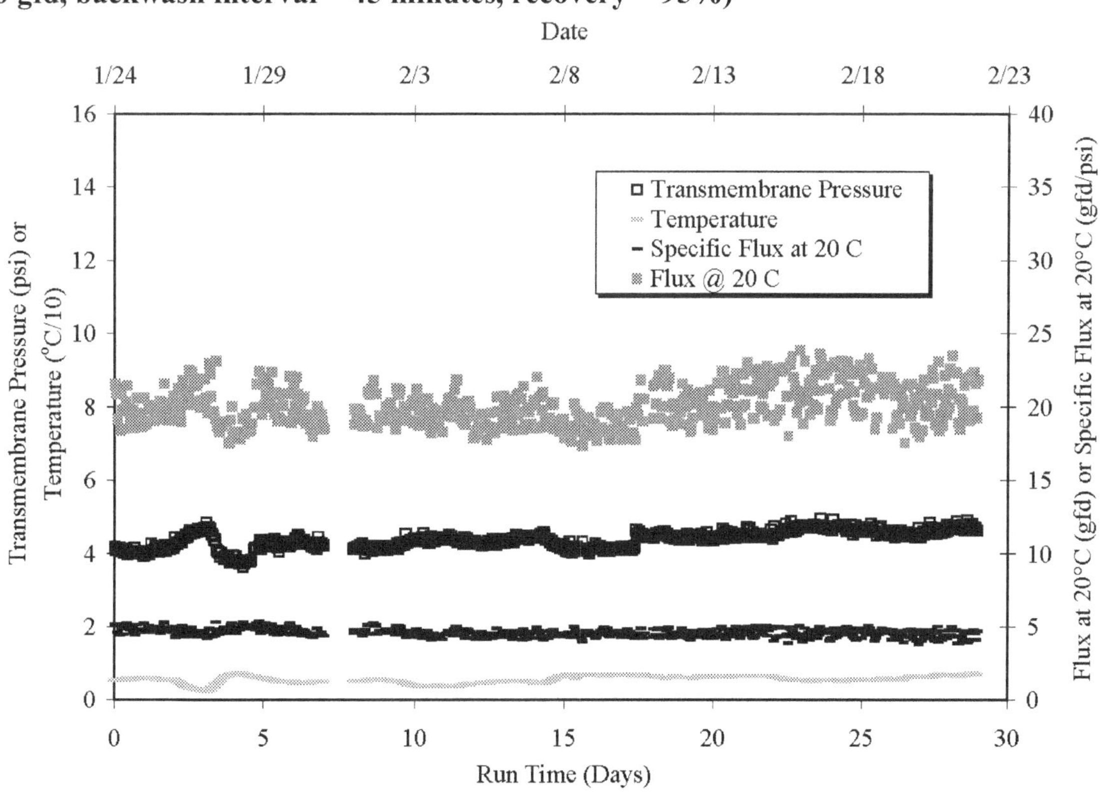

Figure A.10 Run 5 Zenon hydraulic performance (operating conditions: flux @ 20°C = 20.0 gfd, backwash interval = 53 minutes, recovery = 95%)

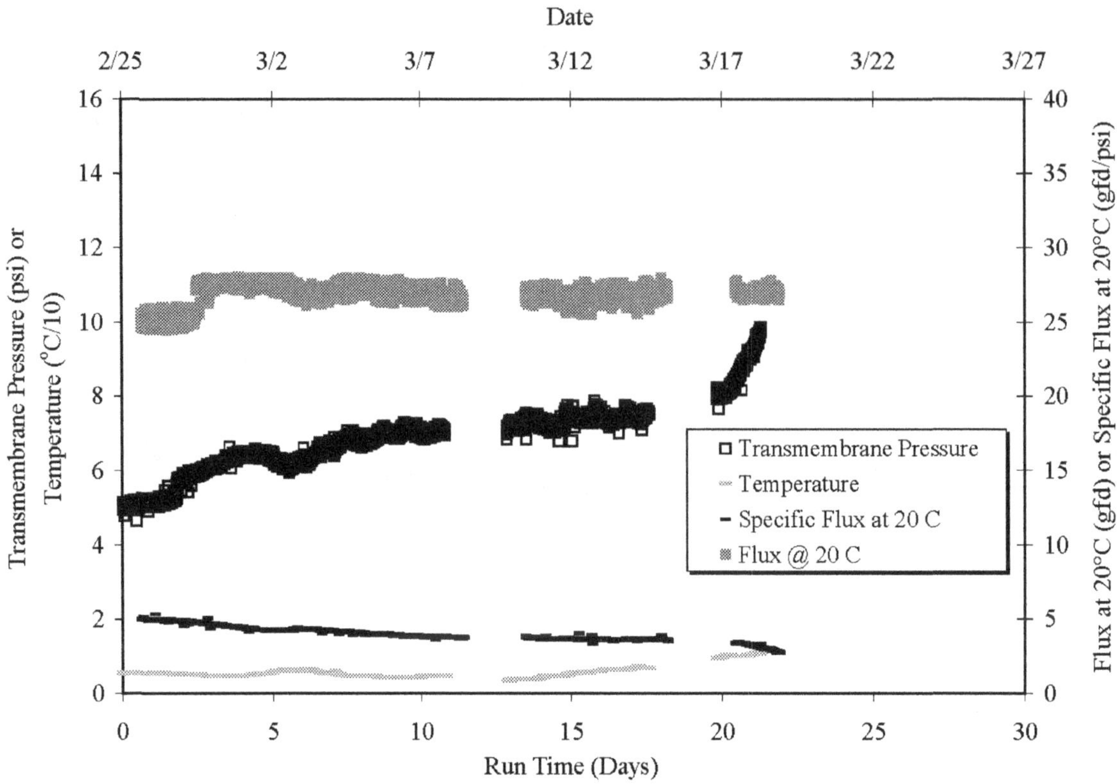

Figure A.11 Run 6 Zenon hydraulic performance (operating conditions: flux @ 20°C = 27.0 gfd, backwash interval = 33 minutes, recovery = 95%)

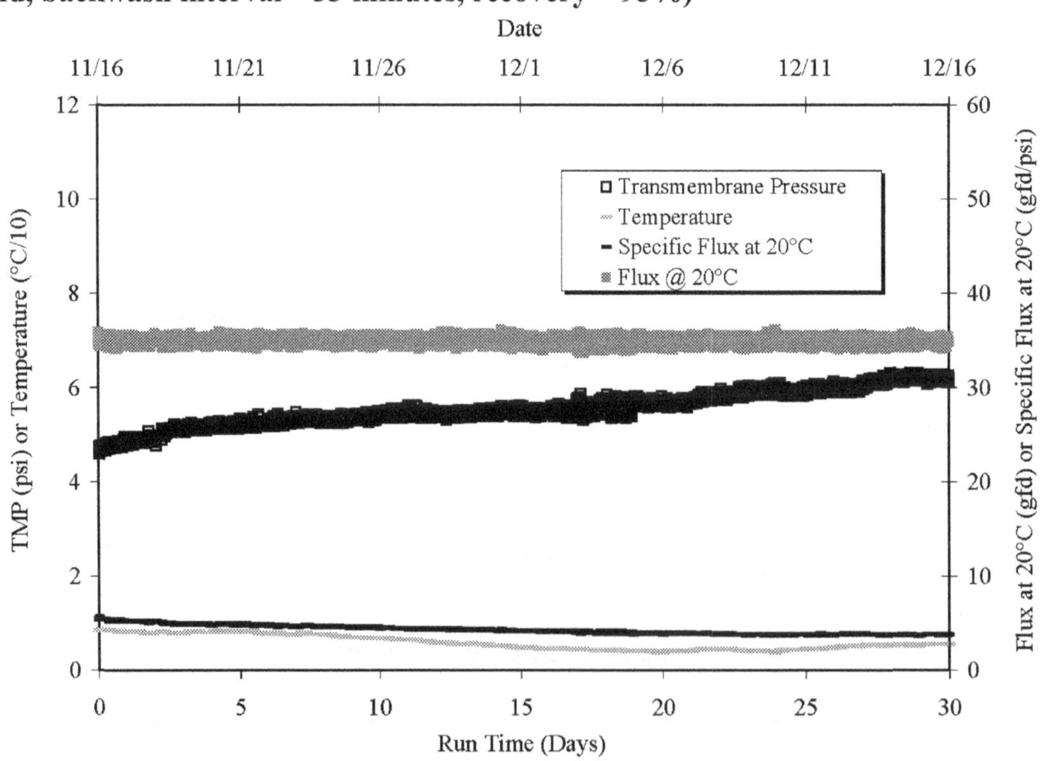

Figure A.12 Run 2 Westch/Polymem hydraulic performance (operating conditions: flux @ 20°C = 35.3 gfd, backwash interval = 34 minutes, recovery = 90%)

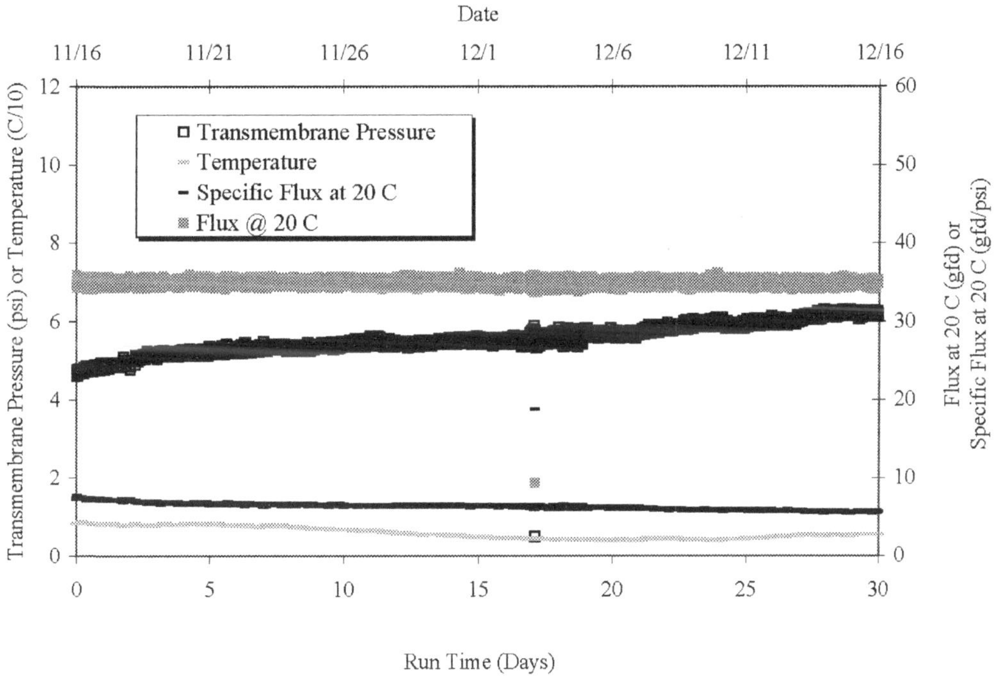

Figure A.13 Run 3 Westech/Polymem hydraulic performance (operating conditions: flux @ 20°C = 41.2 gfd, backwash interval = 33 minutes, recovery = 90%)

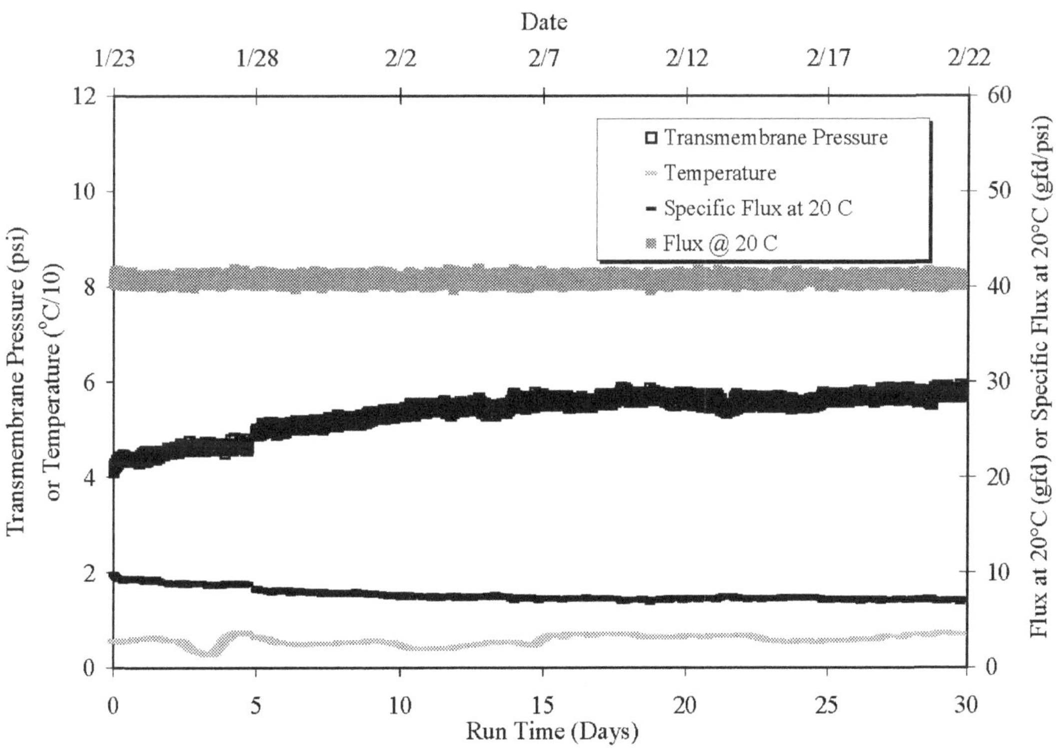

Figure A.14 Run 4 Westch/Polymem hydraulic performance (operating conditions: flux @ 20°C = 41.2 gfd, backwash interval = 39 minutes, recovery = 90%)

Table A.1
INDECK/Hydranautics water quality data – (Run 2: 10/10/02-11/7/02)

Parameter	Feed Water				Permeate Water				Backwash Water			
	ave.	max.	min.	#*	ave.	max.	min.	#	ave.	max.	min.	#
Temperature (°C)	12.5	18.5	7.8	C†	-	-	-	0	-	-	-	0
pH	10.4	10.4	10.4	2	10.2	10.3	10.1	4	10.2	10.3	10.0	4
TSS (mg/L)	31	35	29	9	1	1	1	1	635	960	430	2
Total alkalinity (mg/L as $CaCO_3$)	42	48	34	4	33	33	33	1	35	35	35	1
Total hardness (mg/L as $CaCO_3$)	101	125	92	5	98	98	98	1	-	-	-	0
TOC (mg/L)	2.0	2.2	1.9	3	2.0	2.2	1.8	3	2.1	2.1	2.1	1
DOC (mg/L)	2.0	2.2	1.9	3	1.9	2.1	1.8	3	2.0	2.0	2.0	1
UV_{254} absorption (cm^{-1})	0.047	0.061	0.040	4	0.046	0.048	0.040	4	0.044	0.044	0.044	1
SUVA (L/mg•cm)§	2.4	-	-	3	-	-	-	0	-	-	-	0
Total chlorine (mg/L)	2.4	2.8	2.0	5	-	-	-	0	2.5	2.7	2.3	4
HPC (CFU/mL)	4	8	1	4	5	11	1	5	-	-	-	0
Total coliforms (P/A)‡	A	-	-	4	A	-	-	0	-	-	-	0
Algae (#/100mL)	15	-	-	1	A	-	-	1	-	-	-	0
TTHMs (µg/L)	-	-	-	0	-	-	-	0	-	-	-	0
HAA_5 (µg/L)	13.6	19.9	11.3	4	14.3	19.9	11.0	3	-	-	-	0
[redacted]	% not exceeding			#	% not exceeding			#	ave.	max.	min.	#
	50	95	5		50	95	5					
Turbidity (NTU)	19.8	19.8	15.2	C	0.05	0.06	0.05	C	229	450	76	3
Total particle counts (>2µm; #/mL)	18,000	18,000	18,000	C	0.2	12.6	0.1	C	[redacted]			

* Number of samples
† Continuous measurements
‡ Present/Absent
§ Calculated value

Table A.2
INDECK/Hydranautics water quality data – (Run 3: 11/8/02-12/7/02)

Parameter	Feed water				Permeate water				Backwash water			
	ave.	max.	min.	#*	ave.	max.	min.	#	ave.	max.	min.	#
Temperature (°C)	6.9	10.4	4.5	C†	-	-	-	0	-	-	-	0
pH	10.1	10.3	9.9	3	10.3	10.4	10.1	3	10.3	10.4	10.2	4
TSS (mg/L)	35	64	22	10	-	-	-	0	636	1600	220	2
Total alkalinity (mg/L as $CaCO_3$)	47	56	38	4	39	39	39	1	40	40	40	1
Total hardness (mg/L as $CaCO_3$)	115	131	102	3	100	100	100	1	-	-	-	0
TOC (mg/L)	1.9	2.0	1.8	4	1.9	1.9	1.8	4	2.4	2.4	2.4	1
DOC (mg/L)	1.9	2.0	1.8	4	2.2	2.7	1.9	4	2.8	2.8	2.8	1
UV_{254} absorption (cm^{-1})	0.040	0.044	0.034	4	0.044	0.044	0.044	3	0.036	0.036	0.036	1
SUVA (L/mg•cm)**	2.1	-	-	4	-	-	-	0	-	-	-	0
Total chlorine (mg/L)	2.6	2.7	2.4	4	-	-	-	0	2.6	2.3	2.9	4
HPC (CFU/mL)	1.0	2.2	0.5	4	1.8	2	1	4	-	-	-	0
Total coliforms (P/A)‡	A	-	-	4	A	-	-	0	-	-	-	0
Algae (#/100mL)	-	-	-	0	-	-	-	0	-	-	-	0
TTHMs (µg/L)	-	-	-	0	-	-	-	0	-	-	-	0
HAA_5 (µg/L)	8.4	9.7	7.0	3	7.1	8.4	5.4	3	-	-	-	0
[redacted]	% not exceeding			#	% not exceeding			#	ave.	max.	min.	#
	50	95	5		50	95	5					
Turbidity (NTU)	21.1	28.4	12.1	C	0.06	0.06	0.05	C	583	795	289	3
Total particle counts (>2µm; #/mL)	18,000	18,000	18,000	C	0.02	14.4	0.0	C	[redacted]			

* Number of samples
† Continuous measurements
‡ Did not sample
§ Present/Absent
** Calculated value

Table A.3
INDECK/Hydranautics water quality data – (Run 4: 12/7/02-1/6/03)

Parameter	Feed water				Permeate water				Backwash water			
	ave.	max.	min.	#*	ave.	max.	min.	#	ave.	max.	min.	#
Temperature (°C)	4.0	7.8	2.7	C†	-	-	-	0	-	-	-	0
pH	10.1	10.5	9.9	3	10.0	10.2	9.8	4	10.0	10.1	9.8	4
TSS (mg/L)	25	46	8	8	-	-	-	0	619	880	357	2
Total alkalinity (mg/L as $CaCO_3$)	49	51	41	3	49	49	49	1	53	53	53	1
Total hardness (mg/L as $CaCO_3$)	112	126	103	3	-	-	-	0	1200	1200	1200	1
TOC (mg/L)	1.9	1.9	1.8	3	1.9	2.0	1.9	3	-	-	-	0
DOC (mg/L)	2.1	2.2	1.9	3	2.1	2.3	1.9	3	-	-	-	0
UV_{254} absorption (cm^{-1})	0.050	0.064	0.037	4	0.041	0.066	0.026	4	0.026	0.026	0.026	1
SUVA (L/mg•cm)§	2.4	-	-	3	-	-	-	0	-	-	-	0
Total chlorine (mg/L)	2.4	2.5	2.2	4	-	-	-	0	2.3	2.5	2.2	4
HPC (CFU/mL)	0.9	1.8	0.5	3	3.2	7.0	0.5	3	-	-	-	0
Total coliforms (P/A)‡	A	-	-	4	A	-	-	0	-	-	-	0
Algae (#/100mL)	A	-	-	1	1	-	-	0	-	-	-	0
TTHMs (µg/L)	-	-	-	0	-	-	-	0	-	-	-	0
HAA_5 (µg/L)	-	-	-	0	-	-	-	0	-	-	-	0
[redacted]	% not exceeding			#	% not exceeding			#	Ave.	Max.	Min.	#
	50	95	5		50	95	5					
Turbidity (NTU)	11.1	14.6	9.1	C	0.06	0.07	0.05	C	123	232	71	3
Total particle counts (>2µm; #/mL)	10,123	11,867	7,769	C	0.0	3.5	0.0	C	[redacted]			

* Number of samples
† Continuous measurements
‡ Present/Absent
§ Calculated value

Table A.4
INDECK/Hydranautics water quality data – (Run 5: 1/23/03-2/21/03)

Parameter	Feed water				Permeate water				Backwash water			
	ave.	max.	min.	#*	ave.	max.	min.	#	ave.	max.	min.	#
Temperature (°C)	5.6	7.3	2.5	C†	-	-	-	0	-	-	-	0
pH	10.2	10.7	9.9	3	10.3	10.4	10.1	4	10.3	10.4	10.1	4
TSS (mg/L)	30	44	10	15	18	28	3	7	363	645	81	2
Total alkalinity (mg/L as $CaCO_3$)	57	65	43	4	63	63	63	1	89	89	89	1
Total hardness (mg/L as $CaCO_3$)	126	132	118	4	114	114	114	1	-	-	-	0
TOC (mg/L)	1.6	1.7	1.5	3	1.7	1.9	1.6	4	1.7	1.7	1.7	1
DOC (mg/L)	1.6	1.8	1.5	3	1.8	1.9	1.7	4	1.4	1.4	1.4	1
UV_{254} absorption (cm^{-1})	0.056	0.114	0.031	4	0.073	0.096	0.050	2	0.027	0.027	0.027	1
SUVA (L/mg•cm)§	3.5	-	-	3	-	-	-	0	-	-	-	0
Total chlorine (mg/L)	2.1	2.3	1.9	4	-	-	-	0	1.4	2.7	0.7	4
HPC (CFU/mL)	<1.1	1.8	<0.5	4	1.1	3	0.5	4	-	-	-	0
Total coliforms (P/A)‡	A	-	-	4	A	-	-	4	-	-	-	0
Algae (#/100mL)	A	-	-	1	A	-	-	1	-	-	-	0
TTHMs (µg/L)	-	-	-	0	-	-	-	0	-	-	-	0
HAA_5 (µg/L)	-	-	-	0	-	-	-	0	-	-	-	0
[redacted]	% not exceeding			#	% not exceeding			#	ave.	max.	min.	#
	50	95	5		50	95	5					
Turbidity (NTU)	9.6	13.7	6.9	C	0.05	0.07	0.05	C	225	347	96	4
Total particle counts (>2µm; #/mL)	18,000	18,000	6,990	C	0.0	0.25	0.0	C	[redacted]			

* Number of samples
† Continuous measurements
‡ Present/Absent
§ Calculated value

Table A.5
INDECK/Hydranautics water quality data – (Run 6: 2/25/03-3/30/03)

Parameter	Feed water				Permeate water				Backwash water			
	ave.	max.	min.	#*	ave.	max.	min.	#	ave.	max.	min.	#
Temperature (°C)	8.0	12.2	3.8	C†	-	-	-	0	-	-	-	0
pH	9.8	10.0	9.3	4	9.9	10.2	9.5	5	10.0	10.2	9.7	5
TSS (mg/L)	20	40	2	11	-	-	-	0	1575	1800	1300	2
Total alkalinity (mg/L as $CaCO_3$)	54	64	47	5	52	52	52	1	22	22	22	1
Total hardness (mg/L as $CaCO_3$)	119	143	98	5	116	116	116	1	-	-	-	0
TOC (mg/L)	2.2	2.6	2.0	4	2.3	2.7	2.0	4	2.8	2.8	2.8	1
DOC (mg/L)	2.7	2.7	2.7	1	2.7	2.7	2.7	1	3.0	3.0	3.0	1
UV_{254} absorption (cm^{-1})	0.052	0.061	0.037	5	0.059	0.082	0.035	5	0.060	0.060	0.060	1
SUVA (L/mg•cm)§	1.9	-	-	1	-	-	-	0	-	-	-	0
Total chlorine (mg/L)	2.1	2.5	1.5	5	-	-	-	0	2.2	2.4	1.9	4
HPC (CFU/mL)	<3.8	13.5	0.5	6	<0.5	<0.5	<0.5	4	-	-	-	0
Total coliforms (P/A)‡	A	-	-	4	A	-	-	0	-	-	-	0
Algae (#/100mL)	-	-	-	0	-	-	-	0	-	-	-	0
TTHMs (µg/L)	-	-	-	0	-	-	-	0	9.6	9.6	9.6	1
HAA_5 (µg/L)	-	-	-	0	-	-	-	0	-	-	-	0
	% not exceeding			#	% not exceeding			#	ave.	max.	min.	#
	50	95	5		50	95	5					
Turbidity (NTU)	15.1	25.0	9.3	C	0.06	0.08	0.06	C	415	960	142	4
Total particle counts (>2µm; #/mL)	18,000	18,000	10,019	C	0.1	21.0	0.1	C				

* Number of samples
† Continuous measurements
‡ Present/Absent
§ Calculated value

Table A.6
Zenon water quality data – (Run 2a: 10/10/02-10/21/02)

Parameter	Feed water				Permeate water				Backwash water			
	ave.	max.	min.	#*	ave.	max.	min.	#	ave.	max.	min.	#
Temperature (°C)	15.3	17.3	13.2	C†	-	-	-	0	-	-	-	0
pH	10.4	10.4	10.4	1	10.2	10.2	10.2	1	-	-	-	0
TSS (mg/L)	30	30	30	1	-	-	-	0	1800	1800	1800	1
Total alkalinity (mg/L as $CaCO_3$)	34	34	34	1	-	-	-	0	-	-	-	0
Total hardness (mg/L as $CaCO_3$)	95	98	92	2	96	96	96	1	-	-	-	0
TOC (mg/L)	2.0	2.0	2.0	1	-	-	-	0	-	-	-	0
DOC (mg/L)	2.0	2.0	2.0	1	-	-	-	0	-	-	-	0
UV_{254} absorption (cm^{-1})	0.061	0.061	0.061	1	0.047	0.047	0.047	1	-	-	-	0
SUVA (L/mg•cm)§	3.1	-	-	1	-	-	-	0	-	-	-	0
Total chlorine (mg/L)	2.0	2.0	2.0	1	-	-	-	0	2.2	2.2	2.2	1
HPC (CFU/mL)	8	8	8	1	0.8	1	0.5	2	-	-	-	0
Total coliforms (P/A)‡	A	-	-	4	A	-	-	4	-	-	-	0
Algae (#/100mL)	-	-	-	0	-	-	-	0	-	-	-	0
TTHMs (μg/L)	-	-	-	0	-	-	-	0	-	-	-	0
HAA_5 (μg/L)	-	-	-	0	17.1	17.1	17.1	1	-	-	-	0
	% not exceeding			#	% not exceeding			#	ave.	max.	min.	#
	50	95	5		50	95	5					
Turbidity (NTU)	22.9	25.9	17.6	C	0.04	0.04	0.04	C	-	-	-	0
Total particle counts (>2μm; #/mL)	-	-	-	C	0.7	4.7	0.1	C				

* Number of samples
† Continuous measurements
‡ Present/Absent
§ Calculated value

Table A.7
Zenon water quality data – (Run 2b: 10/25/02-10/31/02)

Parameter	Feed water				Permeate water				Backwash water			
	ave.	max.	min.	#*	ave.	max.	min.	#	ave.	max.	min.	#
Temperature (^oC)	10.0	13.0	9.4	C†	-	-	-	0	-	-	-	0
pH	10.4	10.4	10.4	1	10.1	10.1	10.1	1	10.0	10.0	10.0	1
TSS (mg/L)	31	35	29	3	-	-	-	0	-	-	-	0
Total alkalinity (mg/L as $CaCO_3$)	92	92	92	1	-	-	-	0	-	-	-	0
Total hardness (mg/L as $CaCO_3$)	109	125	92	2	-	-	-	0	-	-	-	0
TOC (mg/L)	2.2	2.2	2.2	1	2.4	2.4	2.4	1	-	-	-	0
DOC (mg/L)	2.2	2.2	2.2	1	2.4	2.4	2.4	1	-	-	-	0
UV_{254} absorption (cm^{-1})	0.041	0.041	0.041	1	0.041	0.041	0.041	1	-	-	-	0
SUVA (L/mg•cm)§	1.9	-	-	1	-	-	-	0	-	-	-	0
Total chlorine (mg/L)	2.7	2.7	2.7	1	-	-	-	0	2.7	2.7	2.7	1
HPC (CFU/mL)	0.8	0.8	0.8	1	-	-	-	0	-	-	-	0
Total coliforms (P/A)‡	A	-	-	4	-	-	-	0	-	-	-	0
Algae (#/100mL)	-	-	-	0	-	-	-	0	-	-	-	0
TTHMs (μg/L)	-	-	-	0	-	-	-	0	-	-	-	0
HAA_5 (μg/L)	11.7	12.0	11.3	2	10.9	11.0	10.7	2	-	-	-	0
[redacted]	% not exceeding			#	% not exceeding			#	ave.	max.	min.	#
	50	95	5		50	95	5					
Turbidity (NTU)	19.0	23.1	14.7	C	0.04	0.04	0.04	C	85	85	85	1
Total particle counts (>2μm; #/mL)	-	-	-	C	0.6	4.3	0.1	C	[redacted]			

* Number of samples
† Continuous measurements
‡ Present/Absent
§ Calculated value

Table A.8
Zenon water quality data – (Run 2c: 11/2/02-11/19/02)

Parameter	Feed water				Permeate water				Backwash water			
	ave.	max.	min.	#*	ave.	max.	min.	#	ave.	max.	min.	#
Temperature (°C)	8.5	9.4	7.1	C†	-	-	-	0	-	-	-	0
pH	10.4	10.4	10.3	2	10.3	10.3	10.3	2	10.2	10.3	10.2	2
TSS (mg/L)	26	37	17	7	-	-	-	0	415	460	370	2
Total alkalinity (mg/L as $CaCO_3$)	46	56	38	4	-	-	-	0	-	-	-	0
Total hardness (mg/L as $CaCO_3$)	101	102	99	2	-	-	-	0	-	-	-	0
TOC (mg/L)	1.9	2.0	1.8	3	1.8	1.9	1.8	1	2.2	2.2	2.2	1
DOC (mg/L)	1.9	2.0	1.8	3	1.8	1.9	1.8	3	2.4	2.4	2.4	1
UV_{254} absorption (cm^{-1})	0.041	0.046	0.034	3	0.041	0.041	0.041	2	-	-	-	0
SUVA (L/mg•cm)§	2.2	-	-	3	-	-	-	0	-	-	-	0
Total chlorine (mg/L)	2.7	2.8	2.6	2	-	-	-	0	2.6	2.6	2.5	2
HPC (CFU/mL)	0.6	0.8	0.5	3	1.5	2	1	2	-	-	-	0
Total coliforms (P/A)‡	A	-	-	4	A	-	-	4	-	-	-	0
Algae (#/100mL)	15	-	-	1	A	-	-	1	-	-	-	0
TTHMs (µg/L)	-	-	-	0	-	-	-	0	-	-	-	0
HAA_5 (µg/L)	10.6	11.4	9.7	2	10.2	10.2	10.2	1	-	-	-	0
[redacted]	% not exceeding			#	% not exceeding			#	ave.	max.	min.	#
	50	95	5		50	95	5					
Turbidity (NTU)	22.8	30.4	16.8	C	0.04	0.04	0.04	C	537	560	513	2
Total particle counts (>2µm; #/mL)	-	-	-	C	1.0	13.7	0.1	C	[redacted]			

* Number of samples
† Continuous measurements
‡ Present/Absent
§ Calculated value

Table A.9
Zenon water quality data – (Run 3: 11/21/02-12/20/02)

Parameter	Feed water				Permeate water				Backwash water			
	ave.	max.	min.	#*	ave.	max.	min.	#	ave.	max.	min.	#
Temperature (°C)	5.6	8.2	3.7	C†	-	-	-	0	-	-	-	0
pH	10.1	10.4	9.9	3	10.2	10.3	10.1	3	10.1	10.2	9.8	3
TSS (mg/L)	35	64	8	8	-	-	-	0	463	550	400	3
Total alkalinity (mg/L as $CaCO_3$)	47	51	42	3	50	50	50	1	-	-	-	0
Total hardness (mg/L as $CaCO_3$)	117	131	108	4	101	101	101	1	530	530	530	1
TOC (mg/L)	1.9	1.9	1.7	3	2.0	2.2	1.9	3	-	-	-	0
DOC (mg/L)	2.0	2.1	1.9	3	2.3	2.9	1.9	3	-	-	-	0
UV_{254} absorption (cm^{-1})	0.050	0.064	0.037	4	0.035	0.042	0.026	3	-	-	-	0
SUVA (L/mg•cm)§	2.0	-	-	3	-	-	-	0	-	-	-	0
Total chlorine (mg/L)	2.4	2.7	2.2	4	-	-	-	0	2.4	2.5	2.3	3
HPC (CFU/mL)	1.1	2.2	0.5	4	3	10	0.5	4	-	-	-	0
Total coliforms (P/A)‡	A	-	-	4	A	-	-	4	-	-	-	0
Algae (#/100mL)	-	-	-	0	-	-	-	0	-	-	-	0
TTHMs (μg/L)	-	-	-	0	-	-	-	0	-	-	-	0
HAA_5 (μg/L)	7.8	8.6	7.0	2	7.9	8.5	6.8	3	-	-	-	0
[redacted]	% not exceeding			#	% not exceeding			#	ave.	max.	min.	#
	50	95	5		50	95	5					
Turbidity (NTU)	11.5	25.2	7.1	C	0.04	0.04	0.04	C	438	549	72	4
Total particle counts (>2μm; #/mL)	-	-	-	C	0.2	8.5	0.1	C	[redacted]			

* Number of samples
† Continuous measurements
‡ Present/Absent
§ Calculated value

Table A.10
Zenon water quality data – (Run 4: 12/22/02-1/20/03)

Parameter	Feed water				Permeate water				Backwash water			
	ave.	max.	min.	#*	ave.	max.	min.	#	ave.	max.	min.	#
Temperature (°C)	4.0	7.8	2.7	C†	-	-	-	0	-	-	-	0
pH	9.9	10	9.9	4	10.2	10.3	10.0	3	10.1	10.3	9.9	3
TSS (mg/L)	27	46	11	11	-	-	-	0	889	1400	487	4
Total alkalinity (mg/L as $CaCO_3$)	48	53	46	2	-	-	-	0	-	-	-	0
Total hardness (mg/L as $CaCO_3$)	135	185	103	4	173	173	173	1	-	-	-	0
TOC (mg/L)	1.9	2.1	1.8	4	2.0	2.4	1.8	4	1.9	1.9	1.9	1
DOC (mg/L)	2.1	2.2	1.9	4	2.1	2.6	1.8	4	2.0	2.0	2.0	1
UV_{254} absorption (cm^{-1})	0.045	0.063	0.037	4	0.040	0.050	0.035	4	0.049	0.049	0.049	1
SUVA (L/mg•cm)§	2.1	-	-	4	-	-	-	0	-	-	-	0
Total chlorine (mg/L)	2.3	2.5	2.2	4	-	-	-	0	2.3	2.5	2.2	4
HPC (CFU/mL)	0.9	1.8	0.5	4	1.4	2.0	0.5	4	-	-	-	0
Total coliforms (P/A)‡	A	-	-	4	A	-	-	4	-	-	-	0
Algae (#/100mL)	A	-	-	1	A	-	-	1	-	-	-	0
TTHMs (μg/L)	-	-	-	0	-	-	-	0	-	-	-	0
HAA_5 (μg/L)	-	-	-	0	-	-	-	0	-	-	-	0
[redacted]	% not exceeding			#	% not exceeding			#	ave.	max.	min.	#
	50	95	5		50	95	5					
Turbidity (NTU)	9.9	14.8	6.5	C	0.04	0.04	0.04	C	238	372	86	3
Total particle counts (>2μm; #/mL)	-	-	-	C	0.2	1.4	0.1	C	[redacted]			

* Number of samples
† Continuous measurements
‡ Present/Absent
§ Calculated value

Table A.11
Zenon water quality data – (Run 5: 1/24/03-2/22/03)

Parameter	Feed water				Permeate water				Backwash water			
	ave.	max.	min.	#*	ave.	max.	min.	#	ave.	max.	min.	#
Temperature (°C)	5.6	7.3	2.5	C†	-	-	-	0	-	-	-	0
pH	10.2	10.7	9.7	3	10.3	10.5	10.1	4	10.2	10.4	9.9	4
TSS (mg/L)	30	44	10	15	20	29	9	7	448	605	290	2
Total alkalinity (mg/L as $CaCO_3$)	57	65	43	4	65	65	65	1	73	73	73	1
Total hardness (mg/L as $CaCO_3$)	126	132	118	4	113	113	113	1	-	-	-	0
TOC (mg/L)	1.6	1.5	1.7	3	1.7	1.6	2.0	4	1.84	1.84	1.84	1
DOC (mg/L)	1.6	1.8	1.5	3	1.9	2.0	1.4	4	1.8	1.8	1.8	1
UV_{254} absorption (cm^{-1})	0.056	0.114	0.031	4	0.060	0.084	0.036	2	0.065	0.065	0.065	1
SUVA (L/mg•cm)§	3.5	-	-	3	-	-	-	0	-	-	-	0
Total chlorine (mg/L)	2.1	2.3	1.9	4	-	-	-	0	1.4	1.9	0.9	4
HPC (CFU/mL)	1.1	1.8	0.5	4	0.6	1	0.5	4	-	-	-	0
Total coliforms (P/A)‡	A	-	-	4	A	-	-	4	-	-	-	0
Algae (#/100mL)	A	-	-	1	A	-	-	1	-	-	-	0
TTHMs (µg/L)	-	-	-	0	-	-	-	0	-	-	-	0
HAA_5 (µg/L)	-	-	-	0	-	-	-	0	-	-	-	0
[redacted]	% not exceeding			#	% not exceeding			#	ave.	max.	min.	#
	50	95	5		50	95	5					
Turbidity (NTU)	8.5	12.0	6.5	C	0.04	0.04	0.04	C	159	368	67	4
Total particle counts (>2µm; #/mL)	-	-	-	C	0.2	4.5	0.1	C	[redacted]			

* Number of samples
† Continuous measurements
‡ Present/Absent
§ Calculated value

Table A.12
Zenon water quality data – (Run 6: 2/25/03-3/18/03)

Parameter	Feed water				Permeate water				Backwash water			
	ave.	max.	min.	#*	ave.	max.	min.	#	ave.	max.	min.	#
Temperature (°C)	8.0	12.2	3.8	C†	-	-	-	0	-	-	-	0
pH	9.7	10.0	9.3	2	10.0	10.2	9.8	3	9.9	10.1	9.7	3
TSS (mg/L)	23	40	4	8	-	-	-	0	340	420	300	3
Total alkalinity (mg/L as $CaCO_3$)	54	64	47	4	52	52	52	1	104	104	104	1
Total hardness (mg/L as $CaCO_3$)	124	143	104	4	115	115	115	1	-	-	-	0
TOC (mg/L)	2.2	2.6	2.0	3	2.4	2.8	2.1	3	2.7	2.7	2.7	1
DOC (mg/L)	2.7	2.7	2.7	1	2.9	2.9	2.9	1	3.4	3.4	3.4	1
UV_{254} absorption (cm^{-1})	0.052	0.061	0.037	4	0.058	0.070	0.040	4	0.064	0.064	0.064	1
SUVA (L/mg•cm)§	1.9	-	-	1	-	-	-	0	-	-	-	0
Total chlorine (mg/L)	2.0	2.5	1.5	4	-	-	-	0	2.2	2.4	1.9	4
HPC (CFU/mL)	<1.0	4.5	0.5	5	<1.0	2.0	<0.5	4	-	-	-	0
Total coliforms (P/A)‡	A	-	-	4	A	-	-	4	-	-	-	0
Algae (#/100mL)	-	-	-	0	-	-	-	0	-	-	-	0
TTHMs (μg/L)	-	-	-	0	-	-	-	0	-	-	-	0
HAA_5 (μg/L)	-	-	-	0	-	-	-	0	-	-	-	0
[redacted]	% not exceeding			#	% not exceeding			#	ave.	max.	min.	#
	50	95	5		50	95	5					
Turbidity (NTU)	13.1	24.8	7.9	C	0.04	0.04	0.04	C	128	167	57	3
Total particle counts (>2μm; #/mL)	18,000	18,000	10,019	C	0.5	11.1	0.2	C	[redacted]			

* Number of samples
† Continuous measurements
‡ Present/Absent
§ Calculated value

Table A.13
Westech/Polymem water quality data – (Run 2: 11/16/02-12/16/02)

Parameter	Feed water				Permeate water				Backwash water			
	ave.	max.	min.	#*	Ave.	max.	min.	#	ave.	max.	min.	#
Temperature (°C)	5.3	8.6	3.9	C†	-	-	-	0	-	-	-	0
pH	10.1	10.4	9.9	3	10.3	10.4	10.1	3	10.1	10.3	9.8	4
TSS (mg/L)	32	64	8	10	-	-	-	0	323	520	160	3
Total alkalinity (mg/L as $CaCO_3$)	46	56	38	4	39	39	39	1	39	39	39	1
Total hardness (mg/L as $CaCO_3$)	117	131	108	3	101	101	101	1	440	440	440	1
TOC (mg/L)	1.9	2.0	1.8	4	2.0	2.1	1.9	3	-	-	-	0
DOC (mg/L)	2.0	2.1	1.9	4	2.2	2.6	2.0	3	-	-	-	0
UV_{254} absorption (cm^{-1})	0.043	0.058	0.034	4	0.039	0.044	0.035	3	0.045	0.045	0.045	1
SUVA (L/mg•cm)§	2.0	-	-	4	-	-	-	0	-	-	-	0
Total chlorine (mg/L)	2.5	2.7	2.2	4	-	-	-	0	3.3	4.2	2.2	4
HPC (CFU/mL)	1.1	2.2	0.5	4	0.6	1.0	0.5	4	-	-	-	0
Total coliforms (P/A)‡	A	-	-	4	A	-	-	4	-	-	-	0
Algae (#/100mL)	-	-	-	0	-	-	-	0	-	-	-	0
TTHMs (μg/L)	-	-	-	0	-	-	-	0	-	-	-	0
HAA_5 (μg/L)	7.8	8.6	7.0	2	7.3	8.0	7.0	2	-	-	-	0
[redacted]	% not exceeding			#	% not exceeding			#	ave.	max.	min.	#
	50	95	5		50	95	5					
Turbidity (NTU)	12.2	22.7	9.3	C	0.04	0.04	0.04	C	175	225	124	4
Total particle counts (>2μm; #/mL)	9,262	10,000	4,956	C	0.3	1.1	0.0	C	[redacted]			

* Number of samples
† Continuous measurements
‡ Present/Absent
§ Calculated value

Table A.14
Westech/Polymem water quality data – (Run 3: 12/20/02-1/20/03)

Parameter	Feed water				Permeate water				Backwash water			
	ave.	max.	min.	#*	ave.	max.	min.	#	ave.	max.	min.	#
Temperature (°C)	4.0	7.8	2.7	C†	-	-	-	0	-	-	-	0
pH	9.9	10.0	9.9	4	10.1	10.3	10.0	3	10.1	10.2	10.1	3
TSS (mg/L)	27	46	11	11	-	-	-	0	251	510	106	4
Total alkalinity (mg/L as $CaCO_3$)	48	53	46	3	63	63	63	1	-	-	-	0
Total hardness (mg/L as $CaCO_3$)	135	185	103	4	155	155	155	1	-	-	-	0
TOC (mg/L)	1.9	2.1	1.8	4	1.9	2.0	1.8	4	1.8	1.8	1.8	1
DOC (mg/L)	2.1	2.2	1.9	4	2.1	2.4	1.8	4	2.0	2.0	2.0	1
UV_{254} absorption (cm^{-1})	0.045	0.063	0.037	4	0.053	0.072	0.035	4	0.041	0.041	0.041	1
SUVA (L/mg•cm)§	2.1	-	-	4	-	-	-	0	-	-	-	0
Total chlorine (mg/L)	2.3	2.2	2.5	4	-	-	-	0	2.9	3.2	2.5	4
HPC (CFU/mL)	0.9	1.8	0.5	4	0.7	1.0	0.5	4	-	-	-	0
Total coliforms (P/A)‡	A	-	-	4	A	-	-	4	-	-	-	0
Algae (#/100mL)	A	-	-	1	A	-	-	1	-	-	-	0
TTHMs (μg/L)	-	-	-	0	-	-	-	0	-	-	-	0
HAA_5 (μg/L)	-	-	-	0	-	-	-	0	-	-	-	0
	% not exceeding			#	% not exceeding			#	ave.	max.	min.	#
	50	95	5		50	95	5					
Turbidity (NTU)	11.1	15.5	8.5	C	0.04	0.04	0.04	C	148	198	98	2
Total particle counts (>2μm; #/mL)	8,032	10,000	3,658	C	0.1	0.3	0.0	C				

* Number of samples
† Continuous measurements
‡ Present/Absent
§ Calculated value

Table A.15
Westech/Polymem water quality data – (Run 4: 1/23/02-2/26/03)

Parameter	Feed water				Permeate water				Backwash water			
	ave.	max.	min.	#*	ave.	max.	min.	#	ave.	max.	min.	#
Temperature (°C)	5.6	7.3	2.5	C†	-	-	-	0	-	-	-	0
pH	10.2	10.7	9.7	3	10.3	10.5	10.1	4	10.2	10.5	9.9	4
TSS (mg/L)	30	44	10	15	-	-	-	0	694	1200	305	4
Total alkalinity (mg/L as $CaCO_3$)	57	65	43	4	63	63	63	1	80	80	80	1
Total hardness (mg/L as $CaCO_3$)	126	132	118	4	114	114	114	1	-	-	-	0
TOC (mg/L)	1.6	1.7	1.5	3	1.7	1.9	1.5	4	1.6	1.6	1.6	1
DOC (mg/L)	1.6	1.8	1.5	3	2.6	2.6	2.6	4	2.9	2.9	2.9	1
UV_{254} absorption (cm^{-1})	0.056	0.114	0.031	4	0.039	0.040	0.039	2	0.041	0.041	0.041	1
SUVA (L/mg•cm)§	3.5	-	-	3	-	-	-	0	-	-	-	0
Total chlorine (mg/L)	2.1	2.3	1.9	4	-	-	-	0	2.1	3.2	1.2	4
HPC (CFU/mL)	<1.1	1.8	<0.5	4	1.0	2	0.5	4	-	-	-	0
Total coliforms (P/A)‡	A	-	-	4	A	-	-	4	-	-	-	0
Algae (#/100mL)	A	-	-	1	A	-	-	1	-	-	-	0
TTHMs (µg/L)	-	-	-	0	-	-	-	0	-	-	-	0
HAA_5 (µg/L)	-	-	-	0	-	-	-	0	-	-	-	0
[redacted]	% not exceeding			#	% not exceeding			#	ave.	max.	min.	#
	50	95	5		50	95	5					
Turbidity (NTU)	8.2	10.9	6.0	C	0.04	0.04	0.04	C	100	143	52	5
Total particle counts (>2µm; #/mL)	10,000	10,000	894	C	0.0	0.5	0.0	C	[redacted]			

* Number of samples
† Continuous measurements
‡ Present/Absent
§ Calculated value

Table A.16
Westech/Polymem water quality data – (Run 5: 2/25/03-3/31/03)

Parameter	Feed water				Permeate water				Backwash water			
	Ave.	max.	Min.	#*	ave.	max.	min.	#	ave.	max.	Min.	#
Temperature (°C)	8.0	12.2	3.8	C†	-	-	-	0	-	-	-	0
pH	9.8	10.0	9.3	4	9.9	10.0	9.7	5	9.6	10.1	8.4	5
TSS (mg/L)	20	40	2	11	-	-	-	0	725	880	570	4
Total alkalinity (mg/L as $CaCO_3$)	54	64	57	5	54	54	54	1	25	25	25	1
Total hardness (mg/L as $CaCO_3$)	119	143	98	5	115	115	115	1	-	-	-	0
TOC (mg/L)	2.2	2.6	2.0	4	2.4	2.6	2.1	4	2.7	2.7	2.7	1
DOC (mg/L)	2.7	2.7	2.7	1	2.6	2.6	2.6	1	2.9	2.9	2.9	1
UV_{254} absorption (cm^{-1})	0.052	0.061	0.037	5	0.061	0.078	0.037	5	0.065	0.065	0.065	1
SUVA (L/mg•cm)§	1.9	-	-	1	-	-	-	0	-	-	-	0
Total chlorine (mg/L)	2.1	2.5	1.5	5	-	-	-	0	3.0	3.9	2.4	5
HPC (CFU/mL)	<3.8	13.5	0.5	6	<1.3	3.0	<0.5	4	-	-	-	0
Total coliforms (P/A)‡	A	-	-	4	A	-	-	4	-	-	-	0
Algae (#/100mL)	-	-	-	0	-	-	-	0	-	-	-	0
TTHMs (μg/L)	-	-	-	0	-	-	-	0	10.4	10.4	10.4	1
HAA_5 (μg/L)	-	-	-	0	-	-	-	0	-	-	-	0
[redacted]	% not exceeding			#	% not exceeding			#	ave.	max.	min.	#
	50	95	5		50	95	5					
Turbidity (NTU)	12.1	19.6	7.4	C	0.04	0.04	0.04	C	103	250	12	4
Total particle counts (>2μm; #/mL)	10,000	10,000	5,014	C	0.8	1.8	0.0	C	[redacted]			

* Number of samples
† Continuous measurements
‡ Present/Absent
§ Calculated value

REFERENCES

Adham, S.S., J.G. Jacangelo, and J.M. Lâiné. 1996. Characteristics and Costs of MF and UF Plants. *Jour. AWWA*, 88(5): 22-31.

Adham, S.S., V.L. Snoeyink, M.M. Clark, and J.L. Bersillon. 1991. Predicting and Verifying Organics Removal by PAC in an Ultrafiltration System. *Jour. AWWA*, 83(12):81-91.

Chui, K., Adham, S., and K. Gramith. 2003. Characteristics of Full-Scale MF/UF Plants. *In Proc. of the 2003 AWWA Water Quality Technology Conference*. Philadelphia, Pa.

Crozes, G.F. and T.F. Seacord. 2001. *Evaluation of Membrane Technologies for Removal of Atrizine and Other SOCs*. Denver, Colo.: AwwaRF.

Freeman, S., M. Horsley, and A. Hess. 2000. Experiences with Submerged and Encased Microfiltration and Ultrafiltration Membranes for Surface Water Treatment. In *Proc. of the Annual AWWA Conference*. Denver, Colo.:AWWA.

Hydranautics. 2002. Technical Product Specifications HYDRACAP. [Online]. Available: <www.membranes.com>.

INDECK Power Equipment Company and Carollo Engineers, P.C. 2003. Kansas City UF Membrane Filtration System: General Arrangement Gallery C. Unpublished.

Lâiné, J-M., M.M. Clark, and J. Mallevialle. 1990. Ultrafiltration of Lake Water: Effect of Pretreatment and Organic Partitioning, THMFP, and Flux. *Jour. AWWA*, 82(12):82-87.

Lawler, D.F., and J. Kweoń. 2003. *Integrated Water Treatment: Softening and Ultrafiltration*. Denver, Colo.: AwwaRF.

Polymem. 2000. UF120 Modules Operation and Maintenance Manual. Unpublished.

U.S. Environmental Protection Agency. 2003. Membrane Filtration Guidance Manual. Office of Water: Fed. Reg. Document 815-D-03-008.

Water Pollution Control Department, Kansas City, Mo. 1979. Secondary Station Principal Mains and Valves. Drawing No. 13359.

Weisner, M.R., and J-M. Lâiné. 1996. *Coagulation and Membrane Separation: In Water Treatment Membrane Processes*. New York: McGraw Hill.

Westech Engineering and Carollo Engineers, P.C. 2003. Gallery C Membrane Retrofit: Plan and Section Views. Unpublished.

Zenon Environmental. 2003. Preliminary Plot Plan "C" Gallery: Membrane Tanks. Retrofit Drawing No. 2125A2-P-001. Unpublished.

Zenon Environmental. 2003. ZeeWeed® Ultrafiltration Performance Evaluation for Kansas City, Missouri: ZeeWeed® 1000 Pilot Summary Report. Unpublished (March).

ABBREVIATIONS

AMSA	active membrane surface area
AWWA	American Water Works Association
AwwaRF	Awwa Research Foundation
BOD	biochemistry oxygen demand
° C	degrees Celsius
CEU	continuing education units
CIP	clean-in-place
DBP	disinfection by-product
DOC	dissolved organic carbon
EPA	Environmental Protection Agency
ft^2	square feet
gfd	gallons per square foot per day
gfd/psi	gallons per square foot per day per pounds per square inch
gpd	gallons per day
gpm	gallon per minute
gpm/ft^2	gallon per minute per square foot
HAA_5	haloacetic acids
HFS	hydro fluorosilicic acid
HLR	hydraulic loading rates
HPC	heterotrophic plate count
IDSE	Initial Distribution System Evaluation
IESWTR	Interim Enhanced Surface Water Treatment Rule
I/O	inside to outside
IT	integrity testing
KCWSD	Kansas City Water Services District
L	liter
LT2ESWTR	Long Term 2 Enhanced Surface Water Treatment Rule
M	million
m^2	square meter
MF	microfiltration
mgd	million gallons per day
mg/L	milligram per liter
mL	milliliter

mm	millimeter
μg/L	microgram per liter
μm	micrometer
NOM	natural organic matter
NPV	net present value
NTU	nephelometric turbidity unit
O&M	Operations and Maintenance
O/I	outside to in
PAC	powdered activated carbon
PDT	pressure decay test
PFD	Process Flow Diagrams
PLC	Programmable Logic Controller
psi	pound per square inch
scfm	standard cubic foot per second
SDWA	Safe Drinking Water Act
SHMP	sodium hexametaphosphate
SLR	surface loading rate
T&O	taste and odor
TTHM	total trihalomethane
TOC	total organic carbon
TMPs	transmembrane pressure
UF	ultrafiltration
WTP	Water Treatment Plant

IWA Publishing's authorised EU representative for General Product Safety Regulations is Diane D'Arras, 15 rue Duret, 75116 Paris, France, e-mail: safety@iwap.co.uk.

Printed and bound by CPI Group (UK) Ltd, Croydon, CR0 4YY
17/09/2026
02218793-0018